Geoheritage, Geoparks and Geotourism

Conservation and Management Series

Spectacular geo-morphological landscapes and regions with special geological features or mining sites are becoming increasingly recognized as critical areas to protect and conserve for the unique geoscientific aspects they represent and as places to enjoy and learn about the science and history of our planet. More and more national and international stakeholders are engaged in projects related to "Geoheritage", "Geo-conservation", "Geoparks" and "Geotourism"; and are positively influencing the general perception of modern Earth Sciences. Most notably, "Geoparks" have proven to be excellent tools to educate the public about Earth Sciences; and they are also important areas for recreation and significant sustainable economic development through geotourism. In order to develop further the understanding of Earth Sciences in general and to elucidate the importance of Earth Sciences for Society, the "Geoheritage, Geoparks and Geotourism Conservation and Management Series" has been launched together with its sister "GeoGuides" series. Projects developed in partnership with UNESCO, World Heritage and Global Geoparks Networks, IUGS and IGU, as well as with the 'Earth Science Matters' Foundation will be considered for publication. This series aims to provide a place for in-depth presentations of developmental and management issues related to Geoheritage and Geotourism in existing and potential Geoparks. Individually authored monographs as well as edited volumes and conference proceedings are welcome; and this book series is considered to be complementary to the Springer-Journal "Geoheritage".

Károly Németh • Vladyslav Zakharovskyi

Geodiversity of Samoa

An Untapped and Unique Geoheritage Hot Spot in the Southwestern Pacific

Springer

Károly Németh
Nat Center for Earthquakes and Volcanoes
Saudi Geological Survey
Jeddah, Saudi Arabia

Vladyslav Zakharovskyi
Volcanic Risk Solutions
Massey University
Palmerston North, New Zealand

ISSN 2363-765X ISSN 2363-7668 (electronic)
Geoheritage, Geoparks and Geotourism

Conservation and Management Series
ISBN 978-3-031-87219-8 ISBN 978-3-031-87220-4 (eBook)
https://doi.org/10.1007/978-3-031-87220-4

This Springer imprint is published by the registered company Springer Nature Switzerland AG
The registered company address is: Gewerbestrasse 11, 6330 Cham, Switzerland

Preface

Geodiversity measures the variety of geological and geomorphological features in an area (Brilha 2016; Coratza et al. 2018; Gray 2018, 2019, 2021, 2023). It plays a crucial role in nature conservation, sustainable planning, tourism, and education (Bailey et al. 2024; Gray 2024; Matthews et al. 2024; Soldati 2023). Managing geodiversity helps address global issues like climate change. Unlike biodiversity, which has established estimation techniques, geodiversity requires a different approach due to its unique spatio-temporal scale (Schrodt et al. 2024; Stojilković and Gray 2024; Zwoliński et al. 2018). Despite intensive research on biodiversity in Pacific small island nations (Carlos et al. 2008), geodiversity studies are significantly lagging. Some new research is also exploring the potential link between geodiversity and biodiversity (Crisp et al. 2023; Oliveira et al. 2024; Salminen et al. 2023; Stojilkovic et al. 2023; Takaoka 2022; Tukiainen et al. 2023; Vernham et al. 2024; Yu and Yang 2024). The SW Pacific, particularly around the Marianas in Micronesia, is recognized as a biodiversity hotspot within the Polynesia-Micronesia bioregion (Willsey et al. 2020). Recent attempts to objectively measure geodiversity involved advanced technologies, including artificial intelligence (Crisp et al. 2021; Forte et al. 2018; Pál and Albert 2023; Wolniewicz 2024). There is also an attempt to express the geodiversity in large regions and use those data for regional planning and conservation strategies (Polman et al. 2024; Ruban 2017; Toivanen et al. 2024).

Geodiversity studies focus on cataloguing geoheritage elements both in situ (natural environment) and ex situ (museums and collections) (da Silva et al. 2023; Nobre et al. 2024; Van Geert 2019) and include tangible (physically touchable) and intangible (spiritual or cultural traditions) (Bruschi and Cendrero 2005; Goemaere et al. 2021) aspects (Brilha 2016). Since the early twenty-first century, geodiversity research has grown significantly, with some regions being more advanced than others. It is increasingly becoming a key area within conservation frameworks and offers practical, interdisciplinary research opportunities for Earth and environmental scientists (Arias-Díaz et al. 2023; Kubalíková et al. 2023; Migoń and Pijet-Migoń 2023). However, estimating geodiversity can be challenging in regions with limited geological activity or geomorphological diversity, such as areas far from tectonic processes or small islands. The SW Pacific region, including Samoa, is beginning to explore geodiversity research (Németh and Gravis 2022). Traditionally known for its tourism—thanks to coral sand beaches, exotic vegetation, and sunshine—the region now sees geodiversity as vital for enhancing tourism, economic development, urban and rural planning, and cultural preservation. This book provides the first comprehensive summary of geodiversity research in Samoa, examining its geological features from volcanoes to coastal formations. It highlights the importance of sustainable programs involving government, communities, and scientific sectors. The summary provides a globally relevant comparative study to objectively define the island's geodiversity and compare its attributes to other regions. Although the methods are still developing, the results offer a solid foundation for further research. The geodiversity model from Samoa is pertinent to other SW Pacific Island nations. This book examines the tourism sector as an economic field where geodiversity research can aid urban and rural planning, fostering co-development between government and local communities. It also contextualizes geotourism within Samoan culture, providing science-based insights for future collaboration. This book represents the first analysis of Samoa's geological and geomorphological potential to utilize its unspoiled abiotic nature.

Samoa is a Polynesian nation with a recorded cultural history of over 3500 years (Jackmond et al. 2022; Martinsson-Wallin 2007). Located in the central region of the Polynesian Triangle, Samoa lies within Oceania, geographically comprised of Melanesia, Micronesia, and Polynesia (Fig. 1). Despite its small landmass (about 9 million km^2 if we include Australia, but without Australia but having New Zealand and Papua New Guinea part of it, the number would be about 1.3 million km^2, equivalent in size to Peru), Oceania is the largest territorial region on Earth but faces unique development challenges due to its economic and demographic size. Though often seen as tranquil tourist destinations, the region experiences violent political conflicts stemming from colonial history, internal mismanagement, geographical limitations, and misleading development policies from external powers (Murray 2009). This complexity warrants the exploration of previously unconsidered areas to achieve sustainable economic growth and environmental preservation. Understanding the region's geodiversity and its geoheritage is therefore a primary objective.

Polynesian cultural diversity is closely linked to geodiversity. Polynesians migrated from offshore islands in Southeast China through the Philippines and Indonesia, as well as Papua New Guinea. This journey, rich in cultural heritage and legends, is supported by modern gene mapping, which shows that migration to the Southwest Pacific originated from Southeast Asia (Brockway 1983; Chambers and Edinur 2020; Finney 2007; Howard 1967; Ioannidis et al. 2021; Stoneking 2025); however, evidence also shows complicated, even reverse movement as small islands mostly in Micronesia and Melanesia were fully colonized (Addison and Matisoo-Smith 2010; Kirk 1982; Palencia-Madrid et al. 2022; Wilson 2018).

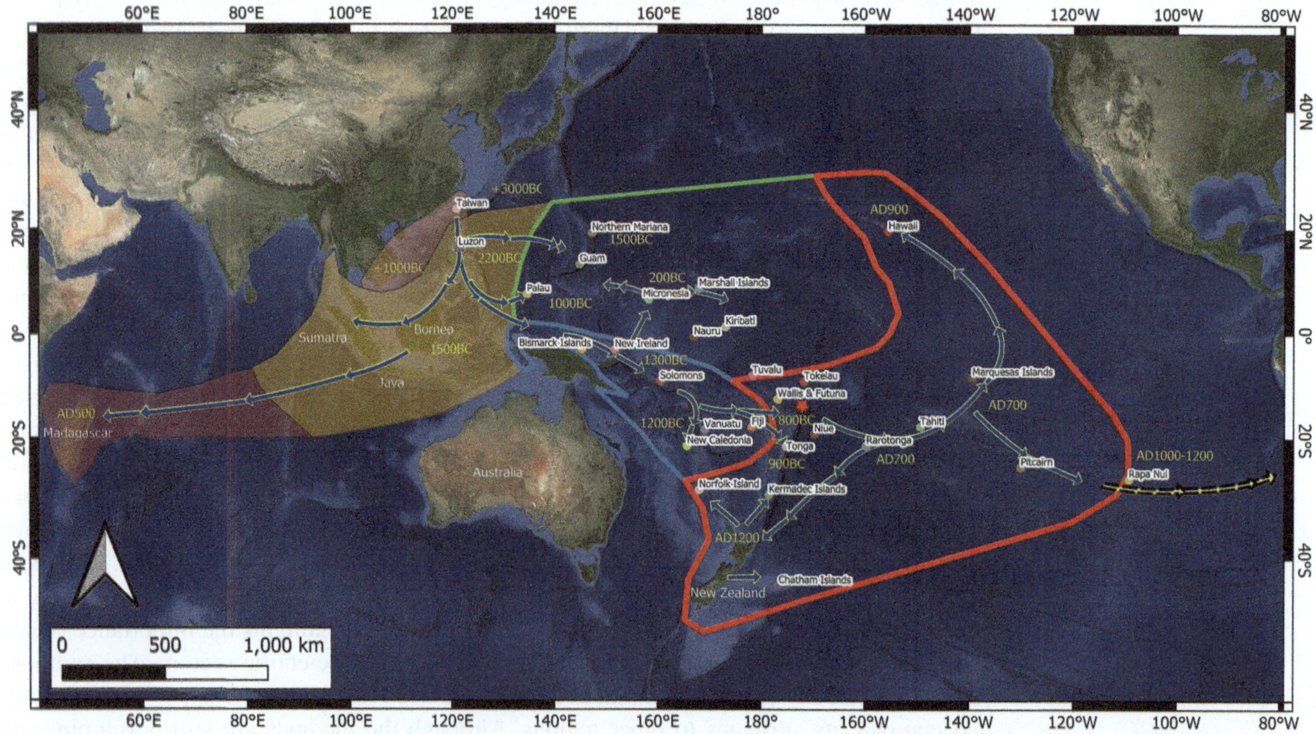

Fig. 1 The human occupation trend based on systematic genetic mapping in combination with archaeological data (Kayser 2010; Stoneking 2025). Red zone outlines Polynesia, green is Micronesia, while blue represents Melanesia. Light pink area around Taiwan identifies as the root of the human migration initiated. Orange fields are SE Asian Island regions, while dark pink region ending with Madagascar shows potential migration path to the west. Polynesia has been reached by about 900 BC and completed by the occupation of New Zealand by AD 1200. Samoa is occupied at least since 800 BC and seems to be a cradle of Polynesian expansion. The SW Pacific of Oceania is commonly distinguished as Remote Oceania (Polynesia and Micronesia together) and Near Oceania (Melanesia with Papua New Guinea and sometimes Australia). This distinction is based on the characteristic biodiversity differences and the geological setting (plate margin versus more intra-oceanic regions) (Green 1991a, b). Dashed yellow line shows the theoretical migration from Rapa Nui to South America

Oceania's human history is marked by three significant migration phases. The first, in the late Pleistocene, saw modern humans move from Southeast Asia to 'Greater Australia' and the Bismarck and Solomon Islands archipelagoes, crossing a vast sea barrier to a land with ancient marsupial fauna. Over 2000 generations later, around 3500 years ago, seafaring pottery-making farmers from Southeast Asia began the 'Lapita phenomenon' in the Bismarck Archipelago and became the first to inhabit Remote Oceania, excluding Antarctica (Lilley 2008). Lapita is recognized as a vast 'community of culture,' which, at least west of Fiji, was sustained by extensive long-distance connections through which people, ideas, and materials traversed significant distances (Burley and Clark 2003; Clark 2009; Donohue and Denham 2012; Green 1991a, b; Kirch and Kahn 2007; Sheppard 2011). The final phase of migration was marked by the arrival of Europeans in Oceania within the last 600 years. We remain in this final phase, as colonialism and its aftermath have promoted ongoing large-scale migration not only into but also within and from Oceania. This has had substantial archaeological consequences and profound social and political implications for archaeological practice, as settlers and indigenous people navigate their differences, many of which are centred on perceptions of the past (Lilley 2008).

Human settlers reached Samoa around 800 BC, approximately 3000 years ago. Before European contact, Samoan culture evolved uniquely, resulting in a rich archaeological landscape and vibrant cultural traditions, including numerous oral traditions. The diverse cultural evolution of Samoa has been extensively explored in recent decades, identifying many aspects considered geocultural heritage. Many elements are intangible, deeply rooted in Polynesian mythology with references to symbiotic living with the environment. This connection is not well articulated, and research directly addressing this is sparse. Although this book does not delve into this complex area, it acknowledges the significant opportunity for future studies to explore the link between indigenous knowledge and the abiotic nature within the Samoan geocultural worldview.

References

Addison DJ, Matisoo-Smith E (2010) Rethinking Polynesians origins: a West-Polynesia Triple-I Model. Archaeol Oceania 45(1):1–12

Arias-Díaz A, Murcia H, Vallejo-Hincapié F, Németh K (2023) Understanding geodiversity for sustainable development in the Chinchiná River Basin, Caldas, Colombia. Land 12(11)

Bailey J, Field R, Schrodt F, van Ree D (2024) Geodiversity for science and society. Philos Trans R Soc A Math Phys Eng Sci 382(2269)

Brilha J (2016) Inventory and qusantitative assessment of geosites and geodiversity sites: a review. Geoheritage 8(2):119–134

Brockway RW (1983) The origin and dispersal of the Polynesians: some recent evidences. J Hum Evol 12(6):501–503

Bruschi VM, Cendrero A (2005) Geosite evaluation: can we measure intangible values. Il Quaternario 18

Burley DV, Clark JT (2003) The archaeology of Fiji/West Polynesia in the post-Lapita era. In: Pacific archaeology: assessments and prospects. Service des Musées et du Patrimonie de Nouvelle-Calédonie, Nouméa, pp 235–254

Carlos G, Velmurugan A, Jerard BA, Karthick R, Jaisankar I (2008) Chapter 4—Biodiversity of Polynesian Islands: distribution and threat from climate change. In: Sivaperuman C, Velmurugan A, Singh AK, Jaisankar I (eds) Biodiversity and climate change adaptation in tropical islands. Academic, New York, pp 105–125

Chambers GK, Edinur HA (2020) Reconstruction of the Austronesian Diaspora in the Era of genomics. Hum Biol 92(4):247–263

Clark G (2009) Post-lapita ceramic change in Fiji. In: The early prehistory of Fiji. ANU Press, Canberra, pp 307–320

Coratza P, Reynard E, Zwoliński Z (2018) Geodiversity and geoheritage: crossing disciplines and approaches. Geoheritage 10(4):525–526

Crisp JRA, Ellison JC, Fischer A (2021) Current trends and future directions in quantitative geodiversity assessment. Prog Phys Geogr 45(4):514–540

Crisp JRA, Ellison JC, Fischer A, Tan JSD (2023) Geodiversity inclusiveness in biodiversity assessment. Prog Phys Geogr 47(3):414–437

da Silva RGP, Mansur KL, de Castro ASFRSF (2023) Methodological proposal for inventory and quantitative valuation of ex situ geological heritage, a case study at the Museu da Geodiversidade (MGeo/IGEO/UFRJ). Geoheritage 15(2)

Donohue M, Denham T (2012) Lapita and proto-oceanic. J Pac History 47(4):443–457

Finney B (2007) Anthropology: tracking Polynesian SEAFARERS. Science 317(5846):1873–1874

Forte JP, Brilha J, Pereira DI, Nolasco M (2018) Kernel density applied to the quantitative assessment of geodiversity. Geoheritage 10(2):205–217

Goemaere E, Millier C, Declercq PY, Fronteau G, Dreesen R (2021) Legends of the Ardennes Massif, a cross-border intangible geo-cultural heritage (Belgium, Luxemburg, France, Germany). Geoheritage 13(2)

Gray M (2018) Geodiversity: the backbone of geoheritage and geoconservation. In: Geoheritage: assessment, protection, and management. Elsevier, Amsterdam, pp 13–25

Gray M (2019) Geodiversity, geoheritage and geoconservation for society. Int J Geoheritage Parks 7(4):226–236

Gray M (2021) Geodiversity: a significant, multi-faceted and evolving, geoscientific paradigm rather than a redundant term. Proc Geologists Assoc 132(5):605–619

Gray M (2023) Some observations and reflections on geodiversity, the oft-forgotten half of nature. Geol Soc Spec Publ 530:13–29

Gray M (2024) Case studies associated with the 10 major geodiversity-related topics. Philos Trans R Soc A Math Phys Eng Sci 382(2269)

Green R (1991a) Near and remote oceania: disestablishing "melanesia" in culture history. In: Pawley A (ed) Man and a half: essays in Pacific Anthropology and ethnobotany in honour of Ralph Bulmer. The Polynesian Society, pp 491–502

Green RC (1991b) The Lapita cultural complex: current evidence and proposed models. Bull Indo-Pacif Prehist Assoc 11:295–305

Howard A (1967) Polynesian origins and migrations. In: Polynesian culture history, pp 45–101

Ioannidis AG, Blanco-Portillo J, Sandoval K, Hagelberg E, Barberena-Jonas C, Hill AVS, Rodríguez-Rodríguez JE, Fox K, Robson K, Haoa-Cardinali S, Quinto-Cortés CD, Miquel-Poblete JF, Auckland K, Parks T, Sofro ASM, Ávila-Arcos MC, Sockell A, Homburger JR, Eng C, Huntsman S, Burchard EG, Gignoux CR, Verdugo RA, Moraga M, Bustamante CD, Mentzer AJ, Moreno-Estrada A (2021) Paths and timings of the peopling of Polynesia inferred from genomic networks. Nature 597(7877):522–526

Jackmond G, Fonoti D, Meleisea M, Schoeffel P, Durling M, Tautunu MM, Sahib M (2022) UTU: Sāmoa archaeology and cultural heritage database. Archaeol Oceania 57(2):95–110

Kayser M (2010) The human genetic history of oceania: near and remote views of dispersal. Curr Biol 20(4):R194–R201

Kirch PV, Kahn JG (2007) Advances in Polynesian prehistory: a review and assessment of the past decade (1993-2004). J Archaeol Res 15(3):191–238

Kirk RL (1982) Microevolution and migration in the Pacific. Prog Clin Biol Res 103 Pt A:215–225

Kubalíková L, Irapta PN, Pál M, Zwoliński Z, Coratza P, Vries BW (2023) Visages of geodiversity and geoheritage: a multidisciplinary approach to valuing, conserving and managing abiotic nature. Geol Soc Spec Publ 530(1):1–12

Lilley I (2008) MIGRATIONS | Pacific. In: Pearsall DM (ed) Encyclopedia of archaeology. Academic Press, New York, pp 1632–1643

Martinsson-Wallin H (2007) Samoan archaeology: a review of research history. Archaeol Oceania 42:11–27

Matthews J, Kubalíková L, Štrba Ľ, Tukiainen H (2024) Geodiversity challenges for a sustainable future. Nat Geosci 17(10):948

Migoń P, Pijet-Migoń E (2023) The role of geodiversity and geoheritage in tourism and local development. Geol Soc Spec Publ 530(1):67–88

Murray WE (2009) Oceania. In: Kitchin R, Thrift N (eds) International encyclopedia of human geography. Elsevier, Oxford, pp 1–13

Németh K, Gravis I (2022) Geoheritage and geodiversity elements of the SW Pacific: a conceptual framework. Int J Geoheritage Parks 10(4):523–545

Nobre AG, Meurer M, Campanaro JF, de Borba AW (2024) Geodiversity elements of geological collections as ex situ teaching tool: the example of the Geography Department at the Federal University of Pelotas, Rio Grande do Sul, Brazil. Geoheritage 16(1)

Oliveira T, Hočevar L, Černe R, Breg Valjavec M, Krofel M (2024) Integrating geodiversity in animal spatial ecology: microhabitat selection of Eurasian lynx (Lynx lynx) and European wildcat (Felis silvestris) in a karst landscape. Global Ecol Conserv 54

Pál M, Albert G (2023) From geodiversity assessment to geosite analysis—a GIS-aided workflow from the Bakony–Balaton UNESCO Global Geopark, Hungary. Geol Soc Spec Publ 530:141–166

Palencia-Madrid L, Baeta M, Kleinbielen T, Toro-Delgado N, Villaescusa P, Sanchez-Bustamante E, de Pancorbo MM, Luis JR, Ware KE, Somarelli JA, Garcia-Bertrand R, Herrera RJ (2022) Post-Austronesian migrational wave of West Polynesians to Micronesia. Gene 823:146357

Polman EMN, Seijmonsbergen AC, Versteegh H, Kissling WD (2024) Global geodiversity components are not equally represented in UNESCO Global Geoparks. Philos Trans R Soc A Math Phys Eng Sci 382(2269)

Ruban DA (2017) Geodiversity as a precious national resource: a note on the role of geoparks. Resour Policy 53:103–108

Salminen H, Tukiainen H, Alahuhta J, Hjort J, Huusko K, Grytnes JA, Pacheco-Riaño LC, Kapfer J, Virtanen R, Maliniemi T (2023) Assessing the relation between geodiversity and species richness in mountain heaths and tundra landscapes. Landsc Ecol 38(9):2227–2240

Schrodt F, Vernham G, Bailey J, Field R, Gordon JE, Gray M, Hjort J, Hoorn C, Hunter ML, Larwood J, Lausch A, Monge-Ganuzas M, Miller S, van Ree D, Seijmonsbergen AC, Zarnetske PL, Kissling WD (2024) The status and future of essential geodiversity variables. Philos Trans R Soc A Math Phys Eng Sci 382(2269)

Sheppard PJ (2011) Lapita colonization across the near/remote oceania boundary. Curr Anthropol 52(6):799–840

Soldati M (2023) Geodiversity and geoheritage research: valuable resources for Earth Science outreach and sustainable development. Geol Soc Spec Publ 530(1)

Stojilković B, Gray M (2024) Geodiversification: the evolution of geodiversity through time. Geoheritage 16(3)

Stojilkovic B, Stepisnik U, Ferk M, Valjavec MB, Cone S (2023) Geodiversity—foundation for life. Geografija v Soli 31(2–3):23–37

Stoneking M (2025) Insights from ancient human DNA into the colonization of Oceania. In: Elias S (ed) Encyclopedia of quaternary science (third edition). Elsevier, Oxford, pp 742–750

Takaoka S (2022) Four types of geodiversity for nature conservation with a focus on the relationship between landform and vegetation. Geoheritage 14(2)

Toivanen M, Maliniemi T, Hjort J, Salminen H, Ala-Hulkko T, Kemppinen J, Karjalainen O, Poturalska A, Kiilunen P, Snåre H, Leppiniemi O, Makopoulou E, Alahuhta J, Tukiainen H (2024) Geodiversity data for Europe. Philos Trans R Soc A Math Phys Eng Sci 382(2269)

Tukiainen H, Toivanen M, Maliniemi T (2023) Geodiversity and biodiversity. Geol Soc Spec Publ 530(1):31–47

Van Geert F (2019) In situ interpretation and ex situ museum display of geology. New opportunities for a geoheritage based dialogue? Int J Geoheritage Parks 7(3):129–144

Vernham G, Bailey J, Field R, Schrodt F (2024) What is the relationship between plant trait diversity and geodiversity? A plot-based, Pan-European analysis. Global Ecol Biogeogr 33(11)

Willsey T, Kwon JA, Reeves MK, Amidon F, Miller SE (2020) Mariana islands forest. In: Goldstein MI, DellaSala DA (eds) Encyclopedia of the world's biomes. Elsevier, Oxford, pp 373–402

Wilson WH (2018) The northern outliers-east Polynesian hypothesis expanded. J Polynesian Soc 127(4):389–423

Wolniewicz P (2024) The combined use of GIS and generative artificial intelligence in detecting potential geodiversity sites and promoting geoheritage. Resources 13(9)

Yu Y, Yang J (2024) Advances in research on the impact of geodiversity on biodiversity. Adv Earth Sci 39(4):347–356

Zwoliński Z, Najwer A, Giardino M (2018) Methods for assessing geodiversity. In: sGeoheritage: assessment, protection, and management. Elsevier, Amsterdam, pp 27–52

Jeddah, Saudi Arabia — Károly Németh
Palmerston North, New Zealand — Vladyslav Zakharovskyi

Contents

1 Samoa in the SW Pacific

The Samoa Islands are in the SW Pacific (Fig. 1.1). Their origin is largely volcanic and record a geological history of about two million of years based on surface or accessible rocks (Fenner 2018; Gale et al. 2021; Goodwin and Grossman 2003; Kear 1967; Kear and Wood 1959). Beside the igneous rocks occur on the islands, sedimentary successions are related with mass wasting such as fluvial and alluvial deposits mostly in the Quaternary. In addition, the offshore regions are characterised with coral reef developments and complex lagune system associated shallow marine sedimentation. The volcanism of the islands is dominated by mafic, mostly basaltic volcanism that are generated by deep upwelling of primary magma from mantle sources through hot spots (Petterson 2020) (Fig. 1.1). The moving of the Pacific Plate roughly toward the west south-west pierced volcanoes through the oceanic crust and formed volcanic islands in the time the melt source was stable and abundant (Jackson et al. 2024; Wei et al. 2024). While the hot spot track and their geochemical signatures are under long lasting debates, the general sentiment is that somehow an eastward migration of volcanism recorded across the island manifested in active volcanism in the most extreme east side of the archipelago (Price et al. 2022). While this trend is roughly recognizable, in fine details there are large number of inconsistencies of age-related trends. The hot spot tracks however in a coarse scale in the SW Pacific perspective showing clearly so-called hot spot highways (Petterson 2020; Price et al. 2022), and Samoa is sitting in one of those zones (Fig. 1.1).

Samoa is also relatively close to a turning point of subduction in the edge of the Pacific Plate (Hrubcová and Vavryčuk 2021) (Fig. 1.1). This means that the archipelago is close to deep marine trench and an area where subduction earthquakes are frequent and capable to generate tsunami waves (Riou et al. 2024; Sischka et al. 2022). The Samoan Island geotectonic situation and their volcanic origin determine their geoheritage assets and their distinction from other Pacific Islands such as many of the older islands of Fiji (Németh and Gravis 2022). The volcanic origin of Samoa also largely dominated by processes that built large lava shield that emerged from the Pacific Ocean and been involved in subsequent erosion and coral reef formation, providing the classical example for atolls. The reef development and the broad shallow marine offshore regions of the oldest and largest islands of the Samoan archipelago provide a unique shallow marine environment with high tropical marine biodiversity and associated geodiversity. While the marine biota is a common subject of research, the sea floor around Samoa has only been partially mapped, and certainly their geodiversity values has not yet been evaluated or appreciated.

The island nature of Samoa also provides a unique geocultural aspect of the region, as it is commonly considered as one of the focal points of the Polynesian culture that is strongly linked to their natural world that has volcanic origin (Németh and Gravis 2022). While in this book we provide some details about the cultural aspect of Samoa, it is not a primary aim to find the link between cultural diversity and geodiversity but acknowledging the importance of this aspect that likely will form subject of future research in the region.

K. Németh, V. Zakharovskyi, *Geodiversity of Samoa*, Geoheritage, Geoparks and Geotourism,
https://doi.org/10.1007/978-3-031-87220-4_1

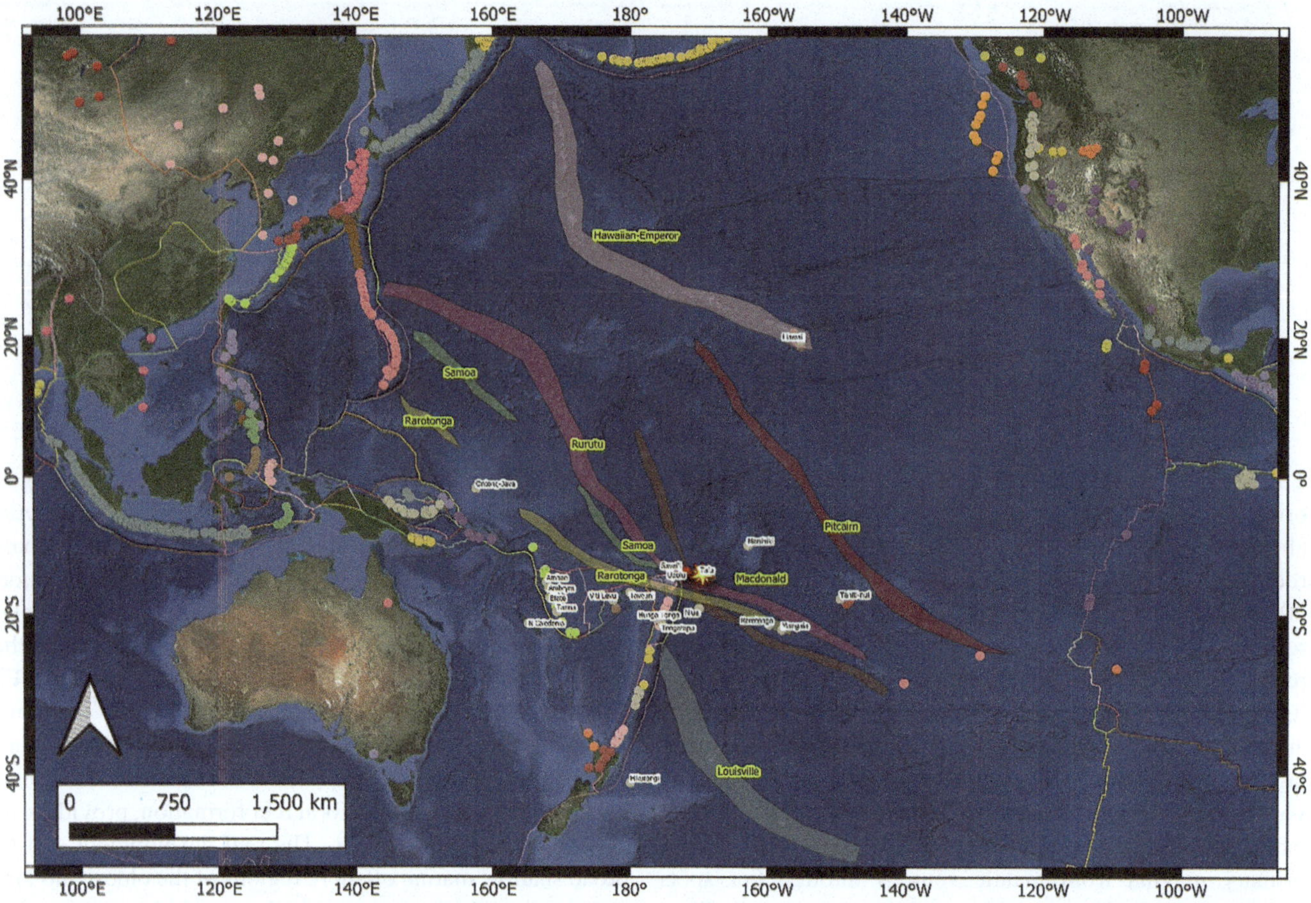

Fig. 1.1 Position of the Samoan Archipelago in the SW Pacific on a Google Earth Pro satellite image. Various coloured zones marking the SW Pacific hot spot highways. Coloured dots showing distinct volcanic zones where Holocene volcanism occurred and formed distinct volcanic zones around the Pacific Rim of Fire. Geotectonic data is from (Bird 2003). Holocene volcano groups are from Volcanoes of the World - Global Volcanism Program https://volcano.si.edu/. Accessed 13 December 2024

1.1 Geodiversity and Geoheritage of Islands

Geodiversity is defined as '*the natural range (diversity) of geological (rocks, minerals, fossils), geomorphological (landforms, topography, physical processes), soil and hydrological features. It includes their assemblages, structures, systems and contributions to landscapes*' (Gray 2013, p. 12). While this definition is the most adapted version of geodiversity, there are great range of interpretation of its meaning in the scientific literature as well as within organizations and individuals working on various enterprises of conservation, planning, education or tourism (Gray et al. 2013; Gray 2018, 2024; Herrera-Franco et al. 2022). Geodiversity research has some flagship countries visible in various research output databases such as Scopus (Fig. 1.2). The main countries most visible in geodiversity research are those where the recognition of the geological and geomorphological diversity is advanced and embraced by various non-profit and government organisations mostly due to pressure from mineral extraction, forestry or just the development and urbanization (Fig. 1.2). Brazil is one of the prominent countries where geodiversity research is advanced and abundant potentially due to the combination of the general diversity issues the country faces with and the strong interest for such research within the research and public sphere (Albani et al. 2020a, b; Alvarenga et al. 2018; Alves et al. 2019; Araujo and Pereira 2018; Arruda et al. 2017; Bento and Rodrigues 2011; Kuhn et al. 2022). On this list only few typical small island countries listed including Jamaica (1), Falkland/Malvinas (1), Bahamas (1), Papua New Guinea (1), French Polynesia (1), Samoa (1), Fiji (1), Faroe Island (1), Dominican Republic (1), Trinidad and Tobago (2), Mauritius (2) and Malta (3) showing very low number of dedicated research outputs on geodiversity from small islands. If larger islands or countries composed of large number of islands (Philippines, Japan, Greece, New Zealand, Madagascar, Ireland or Iceland as best examples), the picture is more promising, but most of the research on those regions associated with much larger

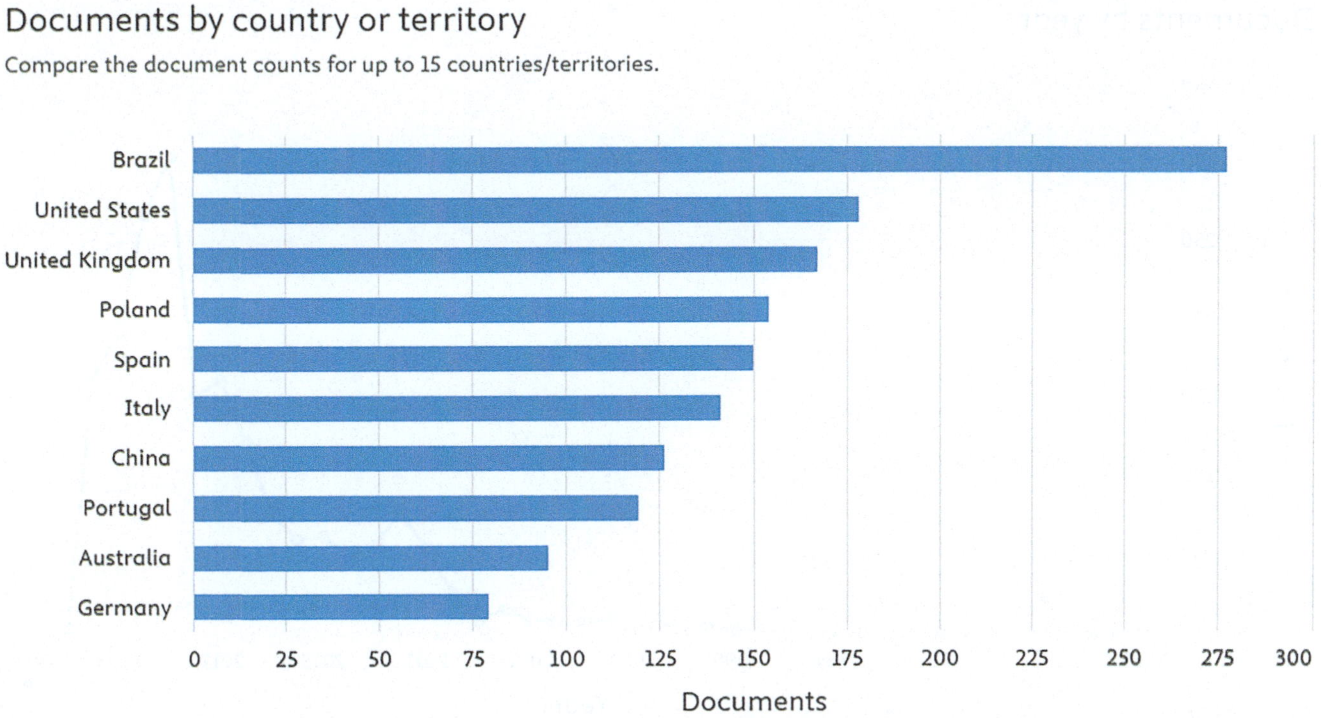

Fig. 1.2 The most common countries as subject of geodiversity research based on Scopus keyword search since the data base hold record (1983) until December 2024

and complex territories as anything in the SW Pacific small islands for instance. From the list there are some exemptions like Spain (15) from where several research papers published from the Canary Islands (Dóniz-Páez and Alonso 2016; Ferrer-Valero 2018; Ferrer-Valero et al. 2019; Páez et al. 2021). Also, as this is not a systematic review (Chen et al. 2020; Giglio et al. 2024; Herrera-Franco et al. 2023; Koupatsiaris and Drinia 2023; Németh and Németh 2023; Ólafsdóttir and Tverijonaite 2018; Quesada-Valverde and Quesada-Román 2023; Zakharovskyi and Németh 2021), it is not considering research outputs published other than English language or Scopus catalogued media.

The common interest in geodiversity research is clearly shown by the exponentially increasing research output (Fig. 1.3).

The author distribution of the geodiversity research in general demonstrates a well-defined group of experts dominate the research field (Fig. 1.4) commonly not from the regions where their research subject is.

Geodiversity commonly mentioned as a sister concept of biodiversity (Crisp et al. 2023; Santos and Mansur 2024; Tukiainen et al. 2023; Vernham et al. 2024; Yu and Yang 2024). While this is a tempting association, geodiversity function in a very different way as biodiversity as it represents a much longer spatio-temporal feature of nature than those biodiversity does (Chen et al. 2024; Stojilković and Gray 2024; Stojilkovic et al. 2023). It is also a more challenging term what to include within and its calculation method also not as well defined as biodiversity (Maliniemi et al. 2024). In recent research the potential to combine geodiversity and biodiversity to express the Earth system diversity also emerged, however the meaning of the results and its practical implication is still long way to understand properly (Crisp et al. 2022). Here in this book, we don't go deep into this discussion.

Geoheritage encompasses global, national, statewide, and local features of geology, at all scales that are intrinsically important sites or culturally important sites offering information or insights into the evolution of the Earth; or into the history of science, or that can be used for research, teaching, or reference (Brocx and Semeniuk 2007). Geoheritage, which has been studied almost since the eighteenth century (Hose 2016), has been rediscovered by the scientific community during the last 25 years. Although much remains to be done before geoheritage is recognised by society and political authorities, great improvements have been made during the last two decades, both at the research and the management levels (Reynard and Brilha 2017). The recognition of what geoheritage is commonly based on the perception that some geodiversity elements have something that it is unusually important, which means that they have an extra value. Due to this high value, these elements must be protected, particularly when the risk of degradation caused by natural or anthropic factors is critical (Brilha 2018). Geoheritage refers to the elements of the Earth that we value,

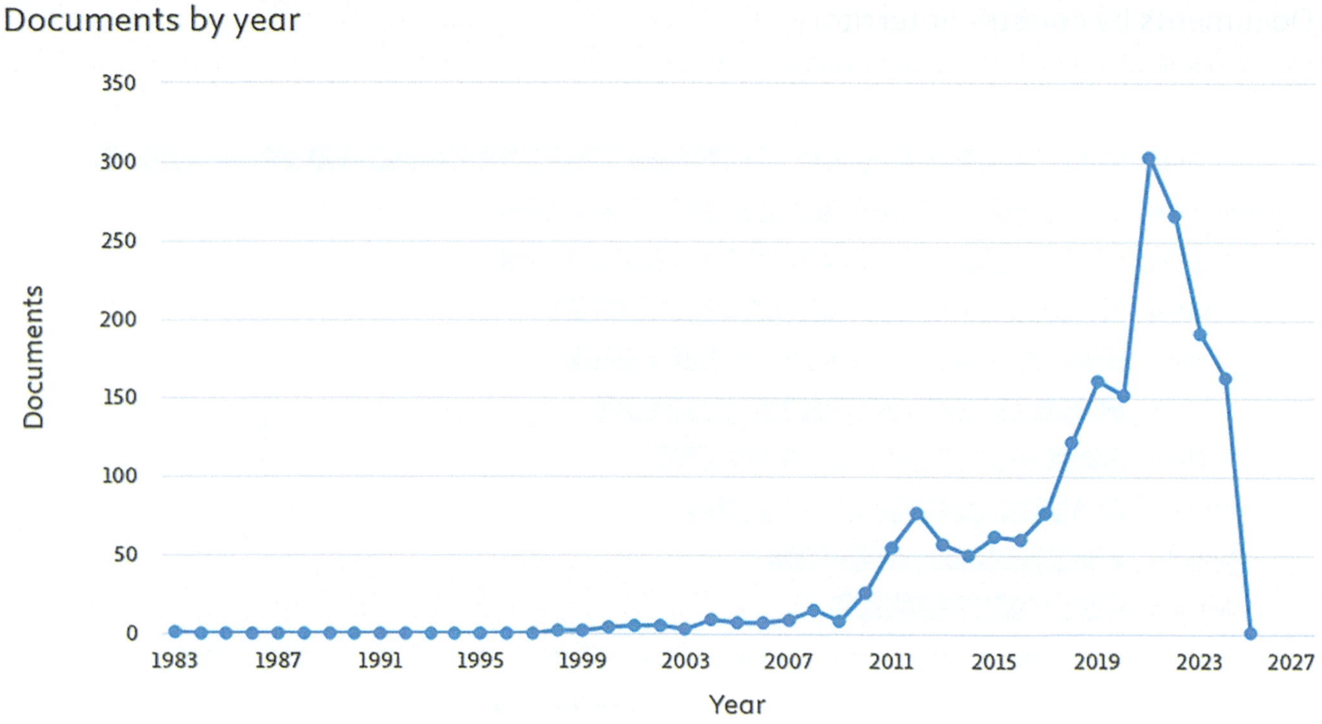

Fig. 1.3 A clear explosion of geodiversity themed research visible in published items per year in the last decade based on Scopus database since 1983 until 2024

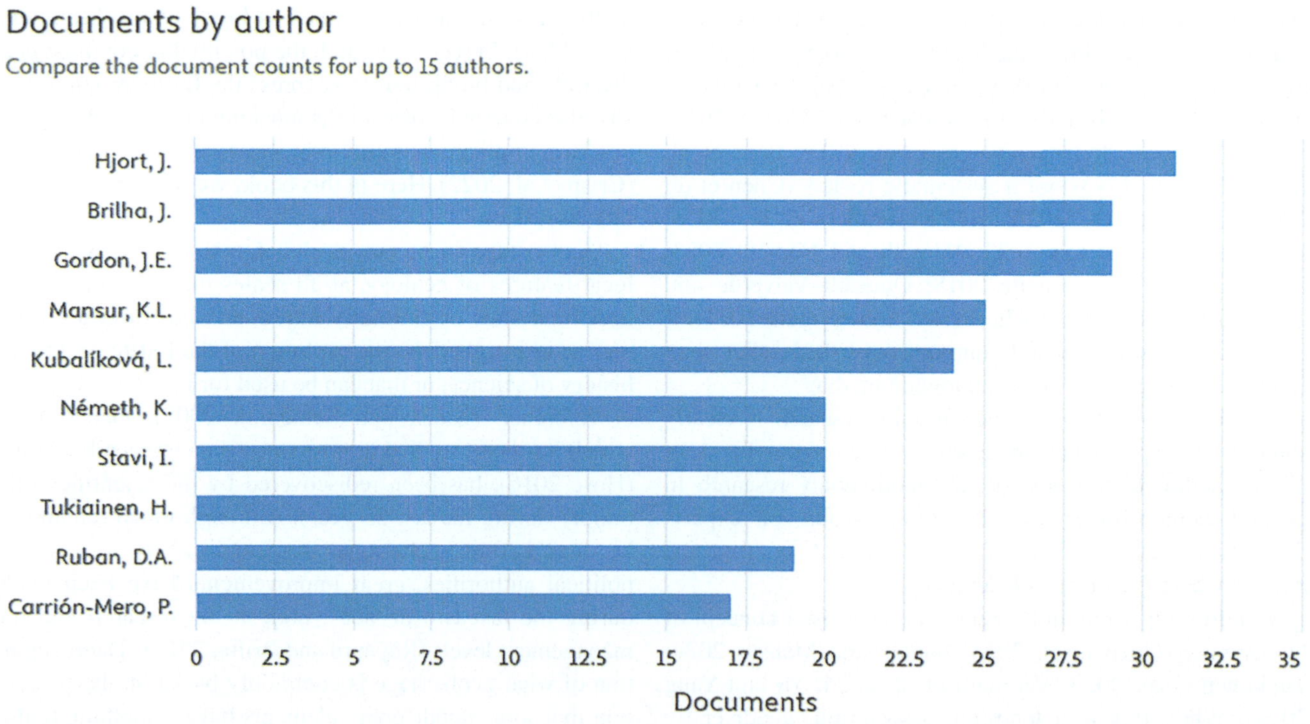

Fig. 1.4 The top 10 authors associated with geodiversity research based on their appearance in citations within Scopus 1983 to 2024 data

whereas geotourism is a type of tourism that is based on some aspects of the Earth's geological and/or geomorphological heritage (Newsome and Dowling 2018). Geotourism may have either positive (beneficial) or negative (adverse) impacts on geoheritage (Newsome and Dowling 2018). At best geotourism fosters an understanding of and a connec-

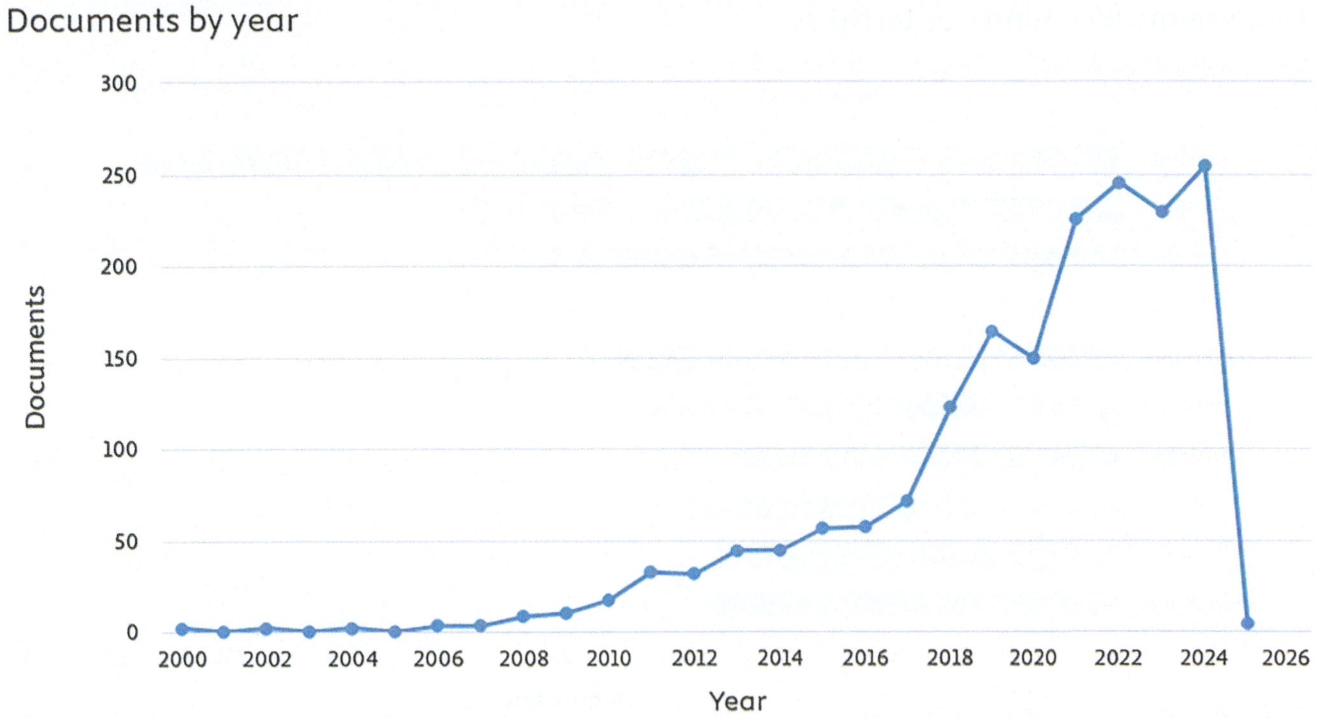

Fig. 1.5 Published items on geoheritage in the Scopus data base between 2000 and 2025 plotted from 1774 documents

tion with geological features through visitation, interpretation and education (Newsome and Dowling 2018).

Exploring the current research trends associated with geoheritage Scopus provides a good data set to analyse. The number of published items (1774) are roughly the same as geodiversity research items (1900) but interestingly the database covers shorter period (2000–2024) (Fig. 1.5). The general trend of the published items appearance is not as dramatically increasing as geodiversity research indicating that the general issue with diversity loss (both bio and geo) became significant issue globally hence research effort increased on those fields.

The geoheritage themed research recorded in Scopus shows different pattern from those tested with geodiversity. While the overlap between countries where geoheritage research is a prominent subject is evident, the ranking is different. This can be interpreted because of the focus of the research as geodiversity research is mostly involved strong spatial statistic support while geoheritage is rather associated with strong traditional geological works. While this issue has not been addressed in scientific research yet, the discrepancy between geodiversity and geoheritage research focusing specific regions is an interesting subject to explore. In the list small island nations are even more underrepresented as having Bahamas (2), Fiji (2), Jamaica (1), Faroe Island (1) and Samoa (1) listed. In this argument larger island nations such as New Zealand, Japan Greece, Cuba, Iceland, Taiwan or Madagascar is not considered. Interestingly, however in this list Philippines is also underrepresented despite its rich geological entity, indicating lack of focused research program to unearth its rich geoheritage.

The researcher list who are associated with geoheritage research and made it to the Scopus list shows also interesting trends (Figs. 1.6 and 1.7).

While these data base analysis important they are not representing the full picture of the state of the research instead showing the way such research reached the global visibility and impact. In local level however geoheritage and geodiversity research may function well and provide vital information to the operation and planning of large regions. This is particularly could be a case in small island nations, such as Samoa, located in isolated areas of the SW Pacific where the academic system is underdeveloped or not expected to function in a same way as in the western world. Still, the geodiversity and geoheritage research of the SW Pacific is largely unknown and non-visible for the rest of the world (Németh and Gravis 2022).

Prior exploring the geoheritage of Samoa, we need to outline the current knowledge of geoheritage within island settings. Relatively low number of research targeted specifically the geoheritage of small islands, and the research visible commonly very specific to a unique geological issue (Ali et al. 2008; Cappadonia et al. 2018; Careddu and Grillo 2019; Corbí et al. 2019; Dóniz-Páez et al. 2020; Fepuleai and Németh 2019; Fepuleai et al. 2021; Galindo et al. 2019; Gioncada et al. 2019; Góis-Marques et al. 2019; Henriques and Neto 2015; Houshold and Sharples 2008; Koh et al. 2014; Mazabraud 2019; Nazaruddin et al. 2016; Páez et al.

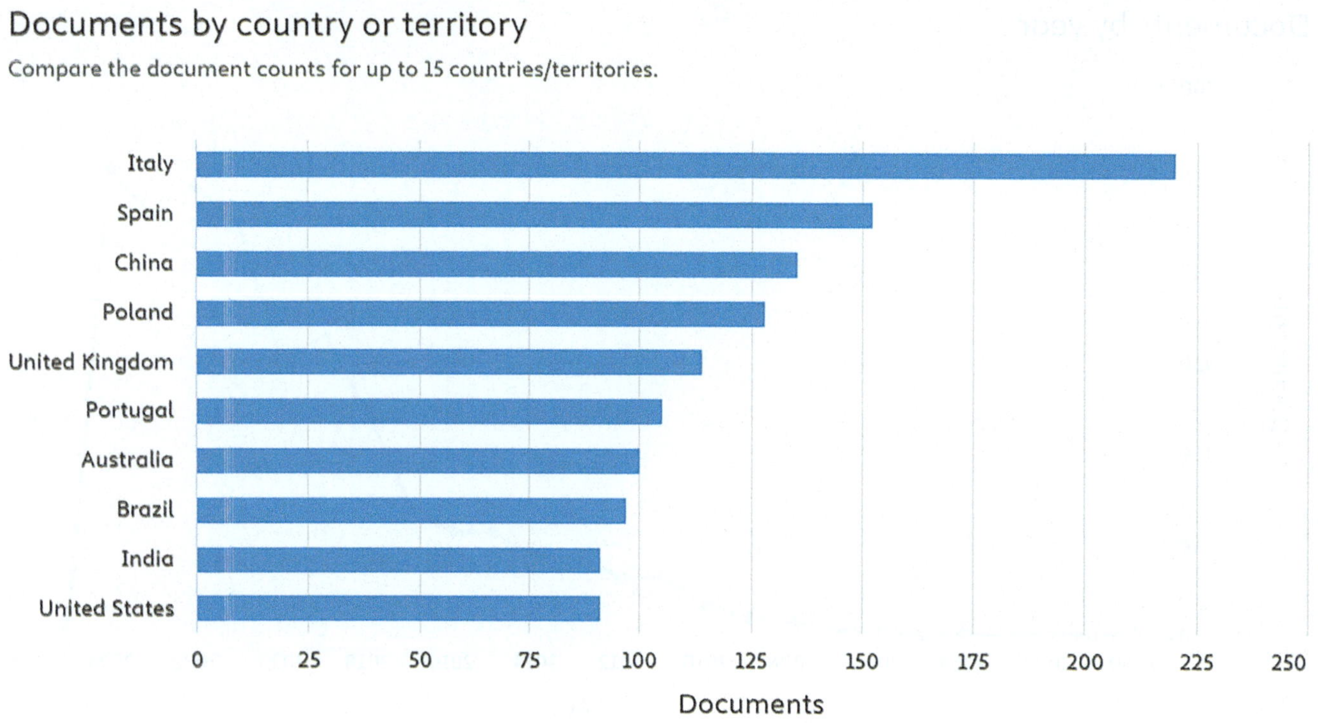

Fig. 1.6 Published items with research subject of geoheritage recorded in the Scopus data base between 2000 and 2025 showing very different pattern from the information on geodiversity

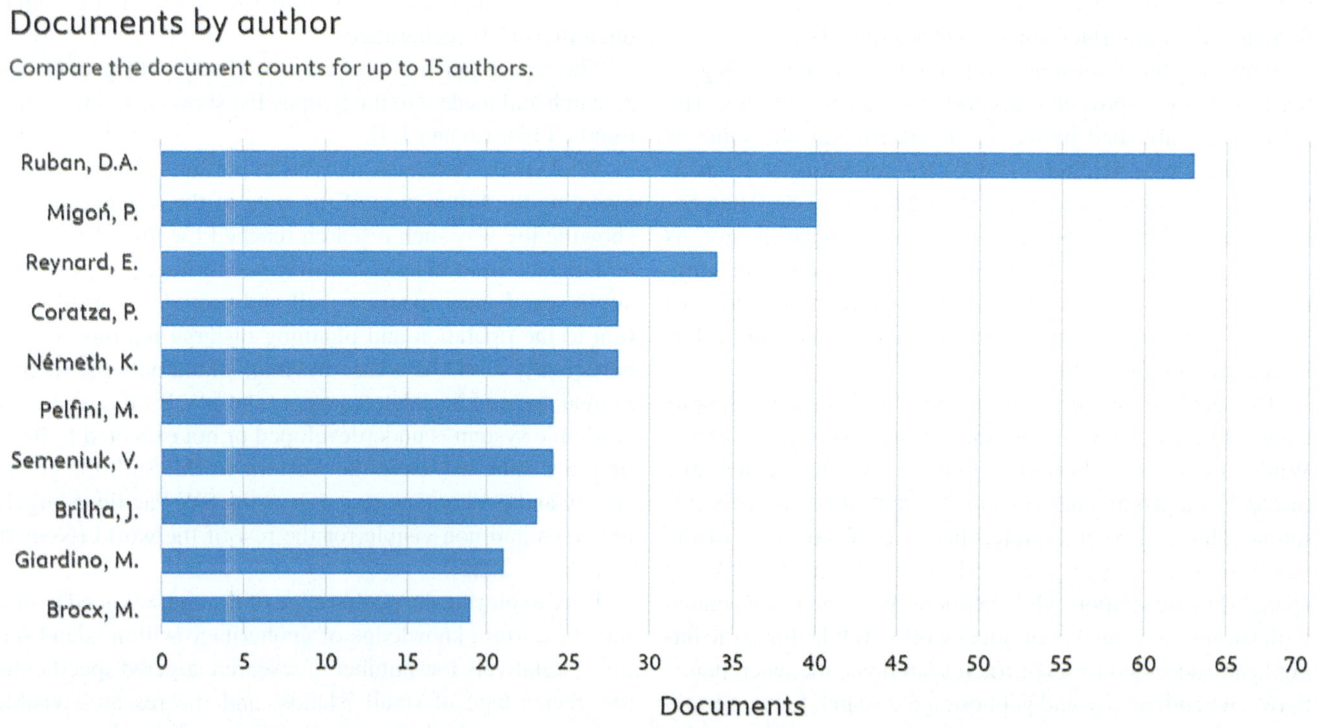

Fig. 1.7 Authors associated with geoheritage research based on the Scopus database between 2000 and 2025

2021; Pagliarulo 2015; Phuong et al. 2013; Raška et al. 2019; Selmi et al. 2019; Woo et al. 2019; Yoon 2019). These examples demonstrate well that the natural diversity of islands is great and the associated researches focusing on the details but less concentrate on potential conceptual framework on how to approach to document geoheritage in island settings.

Islands are special as their landmass is restricted and their geological history normally associated with some special aspects that facilitate their island appearance. Here we refer to only small landmass islands (e.g., typically smaller than 100 km across) and not including large ones like the continent of Australia or even New Zealand as their landmass is significant and their geological history normally associated with processes like other continental settings. The SW Pacific Islands, excluding New Zealand, are the result of complex geotectonical history in the margin of the Pacific Plate and associated break off lithospheric fragments from it, or sandwich between subducting zones (Németh and Gravis 2022). As close we are going to the major convergent plate margins where mature oceanic lithospheric bodies subducting beneath other old oceanic, or continental lithospheric segments, the geological processes showing all the elements associated with the tectonic processes and volcanism associated with typical convergent plate margins. In this respect one of the most complicated geological settings can be noted in the archipelago of the margin of the western Pacific responsible to the formation of the Papua New Guinea, Solomon Islands or Philippines where within very short distances accreted terrains, various carbonate platforms or complex polygenetic volcanoes can form and dominate small island landmasses (Petterson 2020). While the SW Pacific is under the tropical climate zone and abundant rainfall washing their geological features, short rivers with high water yield and rapid erosion and intensive deposition of terrestrial sediments in coastal regions and offshore generate unique geoenvironment that is not visible in such active state elsewhere. This rapid and colourful architecture of the islands in general can make such archipelago a very geodiverse region in a broad scale. However, the scale of few kilometres, or within a single island perspective, there is commonly only specific geological features that are preserved and exposed, providing specific geoheritage to dominate an island. This reduction of potential geodiversity on small islands in the SW Pacific is somehow expected when we move away from the convergent plate margin zones toward the interior of the oceanic plate. Samoa in this scenario is in an area that is not so far from a convergent plate margin, but its landmass still dominated by the result of hot spot volcanism (Fig. 1.1).

The small size of most of the Pacific Islands determines the potential stage where geological processes can take place. In those islands that hosts large stratovolcanoes as the Vanuatu Arc (Fig. 1.1) exposed to rapid mass wasting processes such as landslides that terminate in short distances in the coastal areas without the development of broad coastal fans (Fig. 1.8). Most of the eroded mass end up in the ocean that is quickly deepening. In such setting most of the geological features may carry significant geodiversity values are beneath the water, a realm rarely explored in geoheritage research. The size of the islands certainly the main controlling factors what geodiversity can form on such limited land surface areas (Fig. 1.9). While the small size is a limiting factor, the geological processes responsible for the island growth can significantly impact the type of geoheritage maybe present (Fig. 1.10). In areas where for instance highly explosive vol-

Fig. 1.8 Lopevi Island in the Vanuatu volcanic arc is a single volcanic island of a stratovolcano complex. The island is over 1300 m above sea level and only less than 8 km across generating a very special geological setting with steep slopes on a frequently active volcano. The geodiversity of the island is dominated by the specific mild explosive and effusive volcanic processes providing material to build the island. The coastal region is exposed to high energy wave cut erosion and the low potential to accumulate debris on land while most of material take resting position under the water table (16°28′2.60″S, 168°21′16.04″E looking roughly to the north)

Fig. 1.9 Smaller (less than 5 km across) islands often dominated by a single geoheritage type that is the main geological feature of it. If the geological feature for instance erosion resistance, the island may provide good depositional environment along coastal regions and due to subsidence can host coral reefs. Mataso, Southern Shephard Islands, Vanuatu (17°14′2.13″S, 168°24′41.56″E looking roughly toward the north)

Fig. 1.10 Mayor (Tuhua Island in New Zealand's Bay of Plenty (37°17′7.39″S, 176°17′28.07″E looking toward NW) is an arc volcano that produced silicic lava flows (obsidian), lava domes, and its main morphological elements is defined by a small caldera that produce ash fall reached the mainland of the North Island about 60 km away. The island while it is small, has significant geodiversity due to the variety of explosive and effusive style of volcanism occurred in a relatively short time

canism can take place, small calderas may form, and their associated complex pyroclastic successions may dominate the geomorphosites maybe identified (Fig. 1.10).

The geodiversity of small islands controlled by the type of geological processes occurred in such limited space. In volcanic islands like the Mayor/Tuhua Island in New Zealand, the diverse volcanism determined the diverse volcaniclastic and coherent volcanic rock assemblages accumulated in the Pleistocene to Holocene (Kósik et al. 2022; Zakharovskyi et al. 2023). The variety of rocks accumulated in such small space (less than 5 km across) also very diverse in respect of preservation potential, giving a good basis to have this island great geomorphodiversity (Kósik et al. 2022; Zakharovskyi et al. 2023). The overall geodiversity of the island therefore considerable high (Zakharovskyi et al. 2023).

The complex geodiversity and morphodiversity together with specific climatic and cultural influence can generate islands of small size high geodiversity entity such as those in the Aeolian Islands in Italy (Fig. 1.11).

In the SW Pacific there are many similar examples of islands where active volcanism shaping the islands (Németh and Gravis 2022). If the islands are slightly larger, for instance in the few tens of kilometres across, the specific type of volcanic activity and the total life span of the volcanism can generate very complex landscape with high volcanic geoheritage values such as the Tanna Island in Vanuatu. Yasur, an active volcano of Tanna, made it to the First 100 IUGS Geological Geoheritage Site list (https://iugs-geoheritage.org/geoheritage_sites/the-active-yasur-yenkahe-volcanic-complex-in-vanuatu/—accessed on 14 December 2024) because of its geological significance of the current active stage of a mafic caldera system (Fig. 1.12). Other complex, but frequently active volcanoes in relatively large islands like Ambrym or Ambae in Vanuatu can provide significant amount of sediment in form of various types of pyroclasts can overload the small capacity fluvial network of the islands and generate a sedimentary environment that can be traced from the source to the coastal regions providing textbook type of examples of terrestrial sedimentary systems (Németh et al. 2009) (Fig. 1.12).

The proximity of small islands to each other can add complexity to the geoheritage scene of the islands. Small archipelagos can form complex sedimentary systems in shallow marine environment and in tropical condition can form complex shallow marine geodiversity, and through reef development, a complex biodiversity scene (Fig. 1.13). In islands such as Vanuatu, Solomon Islands in the SW Pacific are fine examples of this type of geodiversity that has hardly ever been researched in the geoheritage or geodiversity context. In these regions, the coastal diversity and the shallow marine seafloor diversity are the key elements of geodiversity. These elements however rarely been conceptualized in geoheritage, or geodiversity perspective and we have very limited tools in hand how to compare, value or look at them from geotourism

Fig. 1.11 The Aeolian Islands in southern Italy are typical high geodiverse small islands. The high geodiversity of the small islands defined by the complex morphodiversity these islands hold due to the diverse rheology the rocks forming the islands (e.g., from rhyolitic lava domes to various intermediate to mafic pyroclasts). The complex eruption styles such as the Vulcanian explosive eruption named from the Island of Vulcano (in front view) is an iconic location that has high scientific values as it is a type of location of a specific eruption style. The islands also have rich geocultural attributes making the small islands a very geodiverse regions (38°24′7.06″N, 14°57′46.15″E looking toward north)

Fig. 1.12 A frequently active mafic volcano such as the Marum and Benbow in the island of Ambrym (Vanuatu9 provides significant amount of ash to the drainage network that carry the sediments quickly to the coastal regions and provide material for the island to grow. This complex sedimentary system within few kilometres distances can provide high sedimentary diversity to the island and scientifically high value subject areas to conduct research to understand the interplay between active volcanism and sedimentary processes (16°16′58.54″S, 168°7′22.48″E looking toward north)

or geoconservation perspective (Avelar et al. 2015; Boretto et al. 2022; Brocx and Semeniuk 2009; Brocx and Semeniuk 2011; Bruno and Perrotta 2012; Coratza et al. 2016; de Arruda et al. 2024; Dóniz-Páez et al. 2020; dos Santos Costa et al. 2024; Fedorov et al. 2019; Fusté-Forné 2023; Guimarães et al. 2017; Gül et al. 2020; Henriques and Neto 2023; Lokier 2013; Possenelli et al. 2024).

In a similarly unexplored field when the geoheritage and geodiversity of small islands are rather located under the water table (Coratza et al. 2019; Huang et al. 2024; O'Leary et al. 2020; Rovere et al. 2011; Vegas et al. 2023). The recent Hunga Tonga–Hunga Ha'apai eruption, that clearly been one of the largest explosive events humanities experienced (Carn et al. 2022; Hu et al. 2023; Lynett et al. 2022; Shrivastava et al. 2023), the main geoheritage and geodiversity elements of the events are either on the seafloor as submerged calderas and volcanic collapse features or scattered and randomly preserved in small islets (Németh 2022) (Fig. 1.13). The preservation potential of most of the eruptive products or the growth of a new islands often hindered by high energy wave cut erosion and the tropical rainfall together, leading the islands vanished rapidly (Fig. 1.13) or destroyed completely by a highly explosive event. The type of volcanism (e.g., hydrovolcanism) and its associated features like distinct clock's tail clouds even inferred as part of the oral traditions of the regions as legends where islands emerged and vanished in the ocean (Fepuleai et al. 2017; Németh and Cronin 2009; Németh and Gravis 2022; Nunn 2001; Nunn 2003). This dynamic landscape and its potential infiltration to cultural traditions provide the strong geocultural aspect of the SW Pacific that also can form as a key element of the geoheritage of the region and its geodiversity (Németh and Gravis 2022). The only issue is, that so far very little attempt has been made to have such elements part of any geodiversity estimates as we just don't have yet a conceptual framework how to include them in a way that can be compared and reproduced objectively.

The Hunga Tonga—Hunga Ha'apai eruption (15 January 2022) highlighted the volcanic hazards the SW Pacific facing as large number of submarine eruptions occurred in the region without significant traces (Németh 2022). Intensified research to locate potential submarine

Fig. 1.13 Quickly eroding tuff cone of the Hunga Tonga volcano prior the enigmatic explosive event in 15 January 2022. The rapid erosion of small islands changing the geoheritage of the island in short period of time (20°32′49.78″S, 175°23′23.80″W, note that the edifice has been demolished by the 15 January 2022 Hunga Tonga—Hunga Ha'apai eruption and tsunami) (Photo by Marco Brenna)

calderas and their deposits in small islets or islands revisited many locations. Also, this enigmatic volcanic events highlighted the problem how to deal with such global and impactful events that leave very limited geological features behind, hence its geoheritage is rather indirectly preserved (Zakharovskyi et al. 2023). Large eruptions, such as the Kuwae eruption that is considered as a climate forcing event in the mid-fifteenth century (its age is under debate, just like its location and potential climate forcing power) associated with submarine calderas (Ballard 2020; Ballard et al. 2023; Gao et al. 2006; Monzier et al. 1994; Németh et al. 2007; Strandberg et al. 2023; Witter and Self 2007). In case of the Kuwae event a seafloor depression between the islands of Epi and Tongoa within the central Vanuatu volcanic arc considered to be a significant and potentially long-lived complex caldera (Fig. 1.14) despite the correlation of its potential deposits and the preservation potential of making any definitive conclusion difficult. This complexity also highlights the problems and pitfalls of geoheritage mapping of islands.

Fig. 1.14 The Kuwae caldera in central Vanuatu is in the seafloor between Tongoa (from where the picture taken) and Epi and some small islets (in the centre right). The regions islands are small, and the rocks preserved there are complex. The geodiversity is high, but their connection is a complex scientific problem highlighting the difficulty to approach geoheritage of small islands (16°50′30.71″S, 168°31′2.23″E)

1.2 Samoa Scene

Samoa consists of two large islands (Figs. 1.15 and 1.16) and several small, mostly volcanic origin (tuff cone) islands along a rift axis. In broader sense the Samoan Archipelago consists of 16 islands with a total surface area of 3126 km^2 and 654 km long coastlines. The archipelago hosts a total of 252,905 inhabitants. Out of this number 186,560 are inhabitants of the two largest islands Upolu (143,418) and Savai'i (43,142) as part of the independent state of Samoa. Tutuila as part of American Samoa just east from Upolu with its 55,876 population is the third populated islands with the third largest surface area. The remaining islands are rather small, less than 100 km^2 in surface areas and population is well below 1000 inhabitants. These small islands are sharing political division as part of the French region of Wallis and Futuna, Tuvalu, American Samoa and Samoa. In this book the geoheritage and geodiversity outline is in Samoa only.

Most of the Samoan islands are in volcanic origin. Volcanic eruptions are known from direct observations, documentations and in many cases in oral legends. The most recent eruption took place in Savai'i, and commonly referred to as the Matavanu eruption that formed an extensive lava field between 1905 and 1911 (Anderson 1910). This location is also became listed in the IUGS First 100 Geological heritage Sites designation (https://iugs-geoheritage.org/geoheritage_sites/1905-1911-matavanu-volcanic-eruption/—accessed on 14 December 2024). Other historic eruptions are known from American Samoa in 1866 in Ofu and Olosega Islands, while in Tutuila an eruption established to occurred in 440 AD (Addison et al. 2006). While the precise eruption chronology of volcanism in the broader Samoan archipelago is generally poorly known, new and new dates gradually emerging, drawing a far more continuous recent volcanism on the islands than it has been thought before. As young, Holocene volcanism is prominent in Samoa, in the next chapter detailed account is given to document those events as highlight their geoheritage and geodiversity significance.

Samoa is in the tropical climate zone that determines its lush vegetation due to high rainfall values reaching about

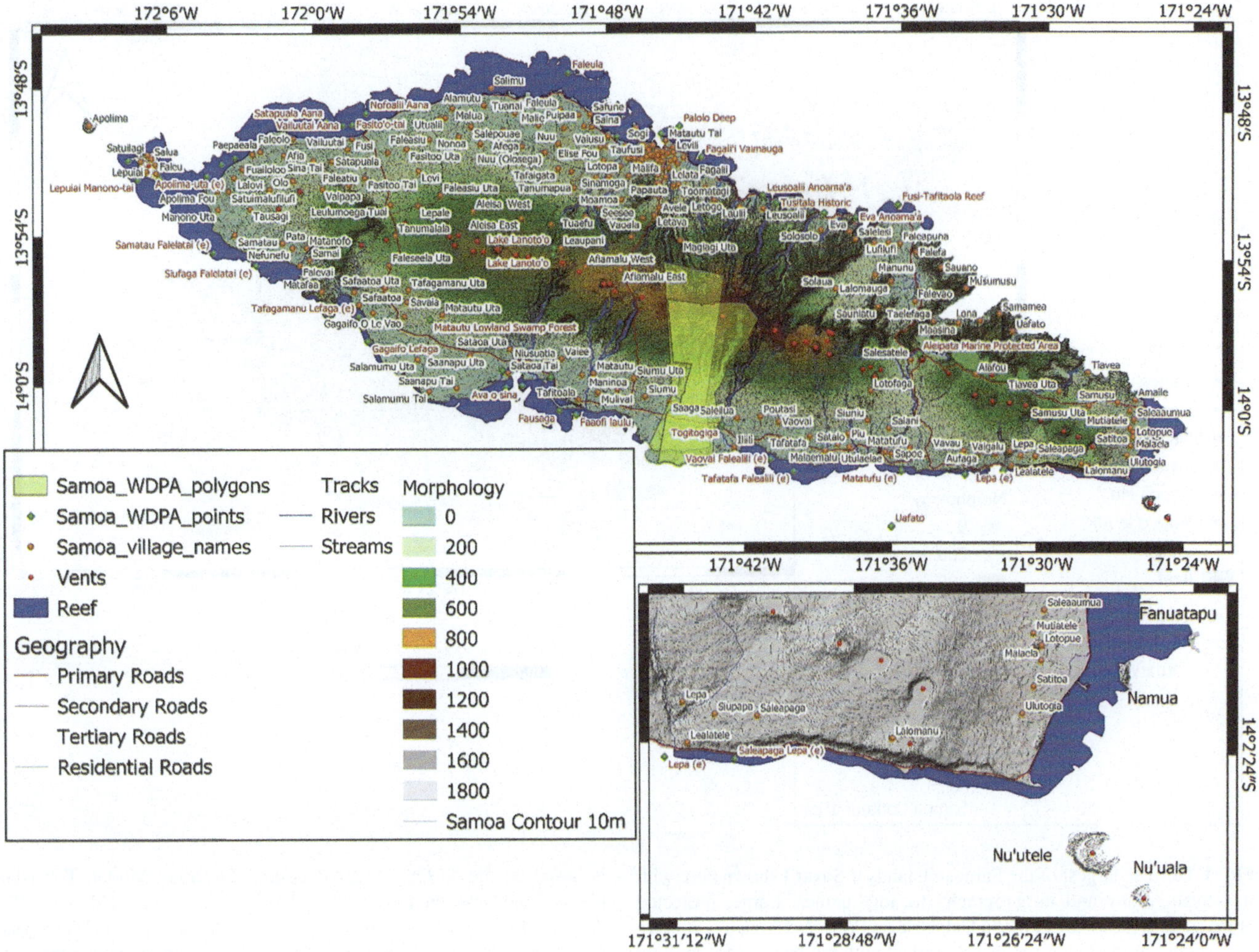

Fig. 1.15 The most populated island, Upolu physiography and its geography locality names. Large, protected areas such as the WDPA (World Database on Protected Areas) also shown. Important reserves are marked with red labels. Morphology is based on the ALOS-PALSAR Digital Elevation Model. The island is heavily populated in its coastal regions, mostly in the secluded reef areas of the NW

3000 mm an average of a year. Rainy seasons span from December to March with an average of monthly fall between 300 and 500 mm. During the driest seasons in June to August, the average rainfall is somewhere between 100 and 150 mm/month. The mean daily minimum and maximum temperatures shows very little variations over the year. The mean daily minimum monthly variations range from about 22–25° while the maximum steadily around 30°. These parameters provide abundant rainwater to feed short stream network and provide vegetation to flourish in the form of tropical rainforest with high biodiversity.

The terrain of the larger islands consists of a narrow coastal plain with volcanic, rocky, rugged mountains in the interior with common deep canyons in the older part of Upolu. The smaller islands are remnants of eroded volcanic tuff cones, some are just a coral reef atop the eroded cone of eroded old volcanoes. The highest point of Samoa is in Savai'i reaching 1858 m above sea level within 20 km from the coast. Upolu is somehow not as tall but still over 1000 m, reaching 1113 m above sea level, within shorter distance from the coast than in the case of Savai'i.

Samoa population living mostly in traditional lands and form village societies. Villages (Figs. 1.15 and 1.16) are largely located along the coastline. Archaeological evidence however showed that before the European contact about 300 years ago, the village network of the islands might have been very different than today, and population centres were in higher grounds. The reason for this is unknown but suspected to be defence from coastal intruders and/or avoiding the humid, commonly malaria-infested wetlands in the coastal regions. The contribution of the rich archaeological record alongside with the islands traditional living culture will be discussed in the geodiversity chapter later.

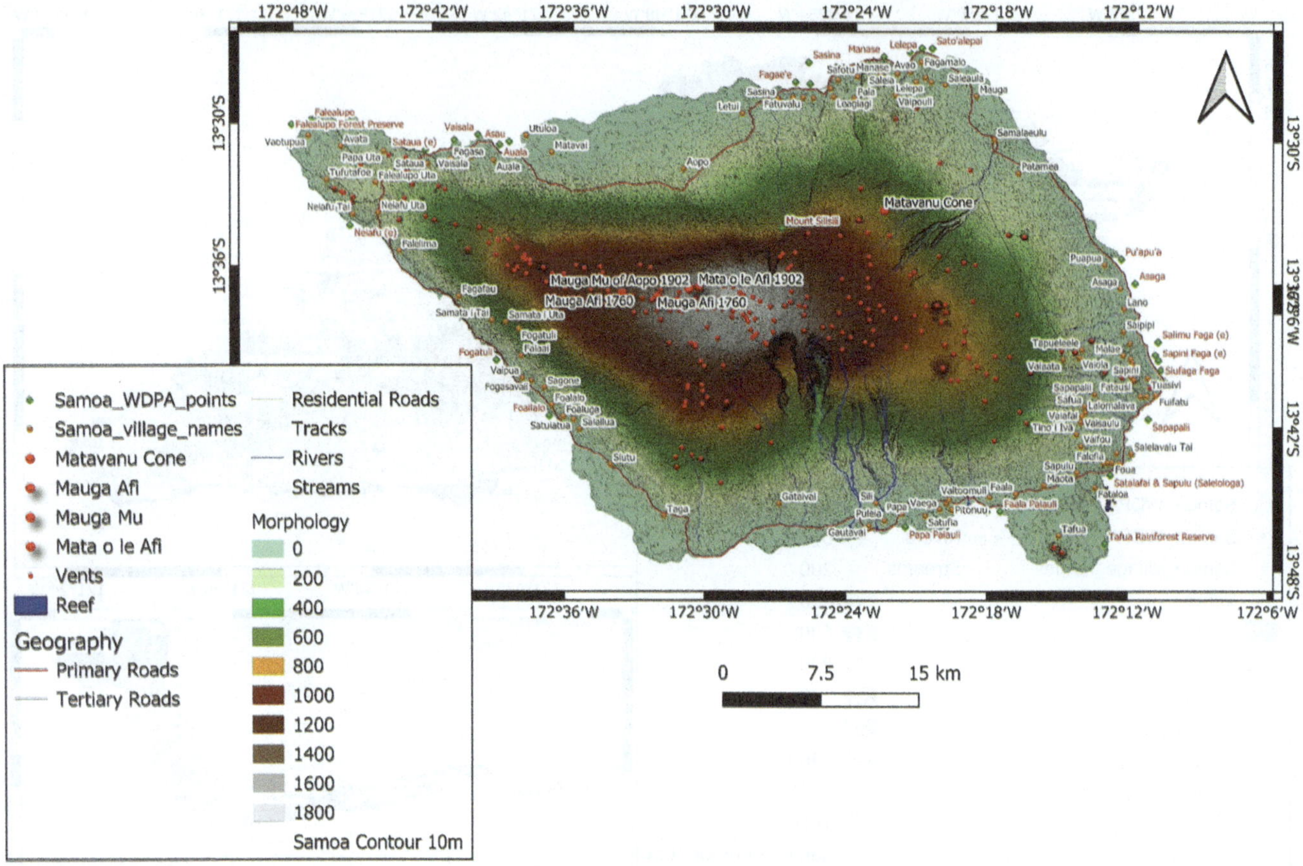

Fig. 1.16 The largest of the Samoan Islands is Savai'I shown here with its physiography and its geography locality names. Large, protected areas such as the WDPA (World Database on Protected Areas) points are also shown. Important reserves are marked with red labels. Morphology is based on the ALOS-PALSAR Digital Elevation Model. The island population is restricted to the coastal regions and dominated in the SE of the island where broad reef provides safe coastal regions. Main volcanoes that were active in Historic time are also marked on the map

1.3 Conclusion

Samoa is a typical small SW Pacific Island group. Geoheritage and geodiversity research has been conducted only on few occasions, and information about the geological architecture of the region largely based on few decades old primary data generation such as mapping. The islands are exclusively volcanic in origin and their formation strongly linked by a within the plate ocean island basalt volcanism. Such volcanism is commonly linked with the lava shield building, erosion, and rejuvenation stage that has been recognized in the Samoa Islands, however the connection between the stages and their volcanism not as clear as it is in Hawaii. Samoa geoheritage hence heavily influenced by volcanism, and in that framework volcanic geoheritage elements are abundant. Samoa geodiversity therefore also determined by the volcanic geoheritage features and all the archaeological and cultural anthropological assets rarely been explored in this context. In this book a summary, a snapshot of Samoa geodiversity and its geoheritage is given with some potential outlook for future research and deeper engagement in global conservation and heritage studies.

References

Addison DJ, Tago T, Toloa J, Pearthree E (2006) Ceramic deposit below fifth to sixth century AD volcanic ash fall at Pava'ia'i, Tutuila Island, American Samoa: preliminary results from site AS-31-171. N Z J Archaeol 27:5–18

Albani RA, Mansur KL, Carvalho IS, Santos WFS (2020a) Quantitative evaluation of the geosites and geodiversity sites of João Dourado Municipality (Bahia—Brazil). Geoheritage 12(2):46

Albani RA, Mansur KL, Sá Dos Santos WF, Pinto ALR (2020b) Beyond sun and beach tourism: a proposal for a geotouristic route for the municipality of São João da Barra. RJ Anu Inst Geocien 43(3):402–414

Ali CA, Mohamed KR, Komoo I (2008) Geoheritage of Pulau Balambangan, Kudat, Sabah. Bull Geol Soc Malaysia 54:91–95

Alvarenga LJ, Castro PTA, Bernardo JM, de Campos IC (2018) Cultural landscape and geoconservation: conceptual contributions applied in Serra da Canastra, Brazil. Anu Inst Geocienc 41(2):241–251

Alves GHZ, Santos RDS, Figueiredo BRS, Manetta GI, Message HJ, Pazianoto LHR, Guimarães GB, Benedito E, Do Couto EV (2019)

Misguided policy may jeopardize a diverse south Brazilian environmental protection area. Biota Neotrop 19(1):1

Anderson T (1910) The volcano of Matavanu in Savaii. Q J Geol Soc 66(1-4):621–639

Araujo AM, Pereira DÍ (2018) A new methodological contribution for the geodiversity assessment: applicability to Ceará State (Brazil). Geoheritage 10(4):591–605

Arruda KE, Garcia MG, Del Lama EA (2017) Inventory and quantitative assessment of geological heritage of the town of Caraguatatuba, São Paulo. Geoheritage 36(3):447–462

Avelar S, Mansur KL, Anjos SC, Vasconcelos GF (2015) Community perceptions for geoconservation of a coastal area in Rio de Janeiro, Brazil. Geoheritage 7(3):275–283

Ballard C (2020) The lizard in the volcano: narratives of the kuwae eruption. Contemp Pac 32(1):98–123

Ballard C, Bedford S, Cronin SJ, Stern S (2023) Evidence at source for the mid-fifteenth century eruption of Kuwae, Vanuatu. J Appl Volcanol 12(1):12

Bento LC, Rodrigues SC (2011) Geodiversity and touristic potential of salto de furnas - Indianópolis-MG. RA'E GA O Esp Geograf Anal 21:272–297

Bird P (2003) An updated digital model of plate boundaries. Geochem Geophys Geosyst 4(3):252

Boretto G, Gordillo S, Cioccale M, Benitez Y, Morán G (2022) Geosite assessment in Northern Patagonia (Argentina): a case study of the coastal geomorphology of the Caleta de Los Loros Natural Protected Area. Geoheritage 14:2

Brilha J (2018) Chapter 4: Geoheritage: inventories and evaluation. In: Reynard E, Brilha J (eds) Geoheritage. Elsevier, Amsterdam, pp 69–85

Brocx M, Semeniuk V (2007) Geoheritage and geoconservation - history, definition, scope and scale. J R Soc West Aust 90:53–87

Brocx M, Semeniuk V (2009) Coastal geoheritage: encompassing physical, chemical, and biological processes, landforms, and other geological features in the coastal zone. J R Soc West Aust 92(3):243–260

Brocx M, Semeniuk V (2011) The global geoheritage significance of the Kimberley coast, Western Australia. J R Soc West Aust 94(2):57–88

Bruno DE, Perrotta P (2012) A geotouristic proposal for Amendolara territory (northern ionic sector of Calabria, Italy). Geoheritage 4(3):139–151

Cappadonia C, Coratza P, Agnesi V, Soldati M (2018) Malta and sicily joined by geoheritage enhancement and geotourism within the framework of land management and development. Geosciences 8(7):253

Careddu N, Grillo SM (2019) "Trachytes" from Sardinia: geoheritage and current use. Sustain For 11(13):3706

Carn SA, Krotkov NA, Fisher BL, Li C (2022) Out of the blue: volcanic SO2 emissions during the 2021–2022 eruptions of Hunga Tonga—Hunga Ha'apai (Tonga). Front Earth Sci 10:976962

Chen SH, Huang J, Tham A (2020) A systematic literature review of coffee and tea tourism. Int J Cult Tour Hosp Res 15(3):290–311

Chen H, Sloggy MR, Escobedo F, Koskimäki T, Lu T, Meng Z, Rasheed AR, Sánchez JJ, Tan X, Yang W, Yu F (2024) Boundary of ecosystem services: differentiating between ecosystem services and geosystem services is needed. J Environ Manag 362:121285

Coratza P, Gauci R, Schembri J, Soldati M, Tonelli C (2016) Bridging natural and cultural values of sites with outstanding scenery: evidence from Gozo, Maltese Islands. Geoheritage 8(1):91–103

Coratza P, Vandelli V, Fiorentini L, Paliaga G, Faccini F (2019) Bridging terrestrial and marine geoheritage: assessing geosites in Portofino Natural Park (Italy). Water 11(10):2112

Corbí H, Martínez-Martínez J, Martin-Rojas I (2019) Linking geological and architectural heritage in a singular geosite: Nueva Tabarca Island (SE Spain). Geoheritage 11(3):703–716

Crisp JRA, Ellison JC, Fischer A (2022) Omnidiversity consolidation of conservation assessment: a case study of Tasmanian coastal geoconservation sites. Geoconserv Res 5(1):108–134

Crisp JRA, Ellison JC, Fischer A, Tan JSD (2023) Geodiversity inclusiveness in biodiversity assessment. Prog Phys Geogr 47(3):414–437

de Arruda IRP, Mariano G, Guimarães TO (2024) Characterization of Pernambuco's geoheritage: analysis of the geodiversity index of the North Coast - Northeast of Brazil. Rev Bras Geogr Fisica 17(1):709–731

Dóniz-Páez J, Alonso CQ (2016) Urban geoturism rutes in icod de los vinos (Tenerife, Canary Islands, Spain): a proposal. Cuad Geograf 55(2):320–343

Dóniz-Páez J, Beltrán-Yanes E, Becerra-Ramírez R, Pérez NM, Hernández PA, Hernández W (2020) Diversity of volcanic geoheritage in the canary islands, Spain. Geosciences 10(10):1–19

dos Santos Costa SS, Junior JCS, Lima ZMC, do Nascimento MA, da Silva ML (2024) Coastal cliffs of the Rio Grande do Norte State: geoheritage characterization and valorization in Northeast Brazil. Geoheritage 16(1):4

Fedorov YA, Gar'kusha DN, Trubnik RG, Latushko NA, Ruban DA (2019) Coastal peloids as geological heritage: evidence from the Taman Peninsula (Southwestern Russia). Water 11(6):1119

Fenner D (2018) The samoan archipelago. In: World seas: an environmental evaluation, vol. II. The Indian Ocean to the Pacific. Elsevier, Amsterdam, pp 619–644

Fepuleai A, Németh K (2019) Volcanic geoheritage of landslides and rockfalls on a tropical ocean island (Western Samoa, SW Pacific). Geoheritage 11(2):577–596

Fepuleai A, Weber E, Németh K, Muliaina T, Iese V (2017) Eruption styles of samoan volcanoes represented in tattooing, language and cultural activities of the indigenous people. Geoheritage 9(3):395–411

Fepuleai A, Németh K, Muliaina T (2021) Geopark impact for the resilience of communities in Samoa, SW Pacific. Geoheritage 13(3):187–193

Ferrer-Valero N (2018) Measuring geomorphological diversity on coastal environments: a new approach to geodiversity. Geomorphology 318:217–229

Ferrer-Valero N, Hernández-Calvento L, Hernández-Cordero AI (2019) Insights of long-term geomorphological evolution of coastal landscapes in hot-spot oceanic islands. Earth Surf Process Landf 44(2):565–580

Fusté-Forné F (2023) A slow tourist in the Basque Coast Geopark (Spain). Int J Geoheritage Parks 11(2):247–258

Gale SJ, Miggins DP, Fepuleai A (2021) Ar-40/Ar-39 dating of quaternary volcanic rocks in Samoa. N Z J Geol Geophys 68:381–388

Galindo I, Romero C, Coello-Bravo JJ, Sánchez N, Martín-González E, Vegas J (2019) A proposal of regional geological frameworks for the geoheritage inventory of the Canary Islands. Geogaceta 65:39–42

Gao C, Robock A, Self S, Witter JB, Steffenson JP, Clausen HB, Siggaard-Andersen ML, Johnsen S, Mayewski PA, Ammann C (2006) The 1452 or 1453 A.D. Kuwae eruption signal derived from multiple ice core records: greatest volcanic sulfate event of the past 700 years. J Geophys Res Atmos 111(12):6710

Giglio VJ, Aued AW, Cordeiro CAMM, Eggertsen L, Ferrari D, Gonçalves LR, Hanazaki N, Luiz OJ, Luza AL, Mendes TC, Pinheiro HT, Segal B, Waechter LS, Bender MG (2024) A global systematic literature review of ecosystem services in reef environments. Environ Manag 73(3):634–645

Gioncada A, Pitzalis E, Cioni R, Fulignati P, Lezzerini M, Mundula F, Funedda A (2019) The volcanic and mining geoheritage of San Pietro Island (Sulcis, Sardinia, Italy): the potential for geosite valorization. Geoheritage 11(4):1567–1581

Góis-Marques CA, Elias RB, Steinbauer M, de Nascimento L, Fernández-Palacios JM, de Sequeira MM, Madeira J (2019) The loss of a unique palaeobotanical site in terceira island within

the azores UNESCO Global Geopark (Portugal). Geoheritage 11(4):1817–1825
Goodwin ID, Grossman EE (2003) Middle to late Holocene coastal evolution along the south coast of Upolu Island, Samoa. Mar Geol 202(1–2):1–16
Gray M (2013) Geodiversity: valuing and conserving abiotic nature. Wiley Blackwell, Chichester
Gray M (2018) Chapter 1: Geodiversity: the backbone of geoheritage and geoconservation. In: Reynard E, Brilha J (eds) Geoheritage. Elsevier, Amsterdam, pp 13–25
Gray M (2024) Case studies associated with the 10 major geodiversity-related topics. Philos Trans R Soc A Math Phys Eng Sci 382:2269
Gray M, Gordon JE, Brown EJ (2013) Geodiversity and the ecosystem approach: the contribution of geoscience in delivering integrated environmental management. Proc Geol Assoc 124(4):659–673
Guimarães T, Mariano G, Barreto A, Sá AA (2017) Beachrocks of Southern Coastal Zone of the State of Pernambuco (Northeastern Brazil): geological resistance with history. Geoheritage 9(1):111–119
Gül M, Küçükuysal C, Çetin E, Ataytür Ö, Masud A (2020) Coastal erosion threat on the Kızkumu Spit Geotourism Site (SW Turkey): natural and anthropogenic factors. Geoheritage 12(3):1–14
Henriques MH, Neto K (2015) Geoheritage at the equator: selected geosites of São Tomé Island (Cameron Line, Central Africa). Sustain For 7(1):648–667
Henriques MH, Neto K (2023) A geo-itinerary to foster sustainable tourism in West African Islands: storytelling the evolution of the ancient Cameroon volcanic line coral reefs. Sustain For 15(24):16863
Herrera-Franco G, Carrión-Mero P, Montalván-Burbano N, Caicedo-Potosí J, Berrezueta E (2022) Geoheritage and geosites: a bibliometric analysis and literature review. Geosciences 12(4):169
Herrera-Franco G, Mata-Perelló J, Caicedo-Potosí J, Carrión-Mero P (2023) Vulnerability in geosites: a systematic literature review. In: World sustainability series. Springer, Cham, pp 395–407
Hose TA (2016) Three centuries (1670–1970) of appreciating physical landscapes. Geological Society, London, pp 1–22
Houshold I, Sharples C (2008) Geodiversity in the wilderness: a brief history of geoconservation in Tasmania. Geol Soc 300:257–272
Hrubcová P, Vavryčuk V (2021) Contact of the Samoan Plume with the Tonga Subduction from intermediate and deep-focus earthquakes. Surv Geophys 42(6):1347–1375
Hu G, Li L, Ren Z, Zhang K (2023) The characteristics of the 2022 Tonga volcanic tsunami in the Pacific Ocean. Nat Hazards Earth Syst Sci 23(2):675–691
Huang Y, He X, Lian Z, Yang Z, Jiang Q (2024) Mapping the landscape of marine cultural heritage research from 2000 to 2023: a bibliometric analysis. Mar Policy 163:106086
Jackson MG, Finlayson VA, Steinberger B, Konrad K (2024) When a plateau suppresses a plume: disappearance of the Samoan Plume under the Ontong Java Plateau. AGU Adv 5(4):e2023AV001079
Kear D (1967) Geological notes on Western Samoa. N Z J Geol Geophys 10(6):1446–1451
Kear D, Wood BL (1959) The geology and hydrology of Western Samoa. N Z Geol Surv Bull 63:1–90
Koh YK, Oh KH, Youn ST, Kim HG (2014) Geodiversity and geotourism utilization of islands: Gwanmae Island of South Korea. J Marine Island Cult 3(2):106–112
Kósik S, Németh K, Rees C (2022) Integrating LiDAR to unravel the volcanic architecture and eruptive history of the peralkaline Tūhua (Mayor Island) volcano. N Z Geomorphol 418:108481
Koupatsiaris AA, Drinia H (2023) Exploring greek UNESCO global geoparks: a systematic review of grey literature on greek universities and future research avenues for sustainable development. Geosciences 13(10):296
Kuhn CES, Santos FRP, de Jesuz CR, de Andrade KA, Vieira AG, Reis F (2022) Public policies for geodiversity in Brazil. Geoheritage 14(2):74
Lokier SW (2013) Coastal Sabkha Preservation in the Arabian Gulf. Geoheritage 5(1):11–22
Lynett P, McCann M, Zhou Z, Renteria W, Borrero J, Greer D, Faanunu O, Bosserelle C, Jaffe B, La Selle SP, Ritchie A, Snyder A, Nasr B, Bott J, Graehl N, Synolakis C, Ebrahimi B, Cinar GE (2022) Diverse tsunamigenesis triggered by the Hunga Tonga-Hunga Ha'apai eruption. Nature 609(7928):728–733
Maliniemi T, Tukiainen H, Hjort J, Toivanen M, Vernham G, Bailey JJ, Baines O, Benniston L, Brilha J, Field R, Fox N, Gray M, Grytnes JA, Huusko K, Kemppinen J, Pereira P, Salminen H, Schrodt F, Turner L, Alahuhta J (2024) Too much diversity—multiple definitions of geodiversity hinder its potential in biodiversity research. Divers Distrib 30(6):e13843
Mazabraud Y (2019) Historical and contemporary use of natural stones in the French West Indies. Conservation aspects and practices. Sustainability 11(17):4566
Monzier M, Robin C, Eissen JP (1994) Kuwae (≈1425 A.D.): the forgotten caldera. J Volcanol Geotherm Res 59(3):207–218
Nazaruddin DA, Mansor HE, Wali SSAS (2016) Geoheritage of labuan island. Bull Geol Soc Malaysia 62:117–129
Németh K (2022) Geoheritage and geodiversity aspects of catastrophic volcanic eruptions: lessons from the 15th of January 2022 Hunga Tonga - Hunga Ha'apai eruption, SW Pacific. Int J Geoheritage Parks 10(4):546–568
Németh K, Cronin SJ (2009) Volcanic structures and oral traditions of volcanism of Western Samoa (SW Pacific) and their implications for hazard education. J Volcanol Geotherm Res 186(3–4):223–237
Németh K, Gravis I (2022) Geoheritage and geodiversity elements of the SW Pacific: a conceptual framework. Int J Geoheritage Parks 10(4):523–545
Németh B, Németh K (2023) Spatial decision-making support for geoheritage conservation in the urban and indigenous environment of the Auckland Volcanic Field. N Z Geol Soc 530(1):235–256
Németh K, Cronin SJ, White JDL (2007) Kuwae caldera and climate confusion. Open Geol J 1:7–11
Németh K, Cronin SJ, Stewart RB, Charley D (2009) Intra- and extra-caldera volcaniclastic facies and geomorphic characteristics of a frequently active mafic island–arc volcano, Ambrym Island, Vanuatu. Sediment Geol 220(3):256–270
Newsome D, Dowling R (2018) Chapter 17: Geoheritage and geotourism. In: Reynard E, Brilha J (eds) Geoheritage. Elsevier, Amsterdam, pp 305–321
Nunn PD (2001) On the convergence of myth and reality: examples from the Pacific Islands. Geogr J 167(2):125–138
Nunn PD (2003) Fished up or thrown down: the geography of Pacific Island origin myths. Ann Assoc Am Geogr 93(2):350–364
O'Leary MJ, Paumard V, Ward I (2020) Exploring sea country through high-resolution 3D seismic imaging of Australia's NW shelf: resolving early coastal landscapes and preservation of underwater cultural heritage. Quat Sci Rev 239:106353
Ólafsdóttir R, Tverijonaite E (2018) Geotourism: a systematic literature review. Geosciences 8(7):234
Páez JD, Ramírez RB, Yanes EB (2021) Geoheritage and geotourism in volcanic tenerife's natural protected areas (Canary islands, Spain). Cuad Geogr 60(2):52–71
Pagliarulo R (2015) Relationship between natural hazards and geological heritage: the case of cretaccio Island (Tremiti Archipelago, Southern Italy). Geoheritage 7(1):57–63
Petterson MG (2020) Geo-tectonics, geology, and geo-resources of the southwest pacific. In: Encyclopedia of geology: volume 1–6, 2nd edn. Elsevier, Amsterdam, pp 724–741
Phuong TH, Cu NH, Thanh TD, Van Dong B (2013) Geoheritage values in the Cat Ba islands. Vietnam Environ Earth Sci 70(2):543–548

Possenelli M, Gauci R, Devoto S, Selmi L, Coratza P, Vandelli V (2024) Inventory and quantitative assessment of geosites in the Southern Sector of the Island of Malta. Geosciences 14(11):292

Price AA, Jackson MG, Blichert-Toft J, Konrad K, Bizimis M, Koppers AAP, Konter JG, Finlayson VA, Sinton JM (2022) Distinguishing volcanic contributions to the overlapping Samoan and Cook-Austral hotspot tracks. J Petrol 63(5):32

Quesada-Valverde ME, Quesada-Román A (2023) Worldwide trends in methods and resources promoting geoconservation, geotourism, and geoheritage. Geosciences 13(2):39

Raška P, Pokorný R, Krmíček L, Kuboušková S, Mortensen L (2019) Basaltic dyke with specific volcanogenic structures and its geomorphic evolution: unique geoheritage of the faroe islands (North Atlantic Ocean). Geoheritage 11(2):417–426

Reynard E, Brilha J (2017) Geoheritage: assessment, protection, and management. Elsevier, Amsterdam, pp 1–450

Riou B, Chaumillon E, Chagué C, Schmidt S, Corrège T, Bujan S, Schneider JL (2024) Offshore evidence of historic and prehistoric tsunamis on the north shore of Tutuila (American Samoa). Sediment Geol 461:106572

Rovere A, Vacchi M, Parravicini V, Bianchi CN, Zouros N, Firpo M (2011) Bringing geoheritage underwater: definitions, methods, and application in two Mediterranean marine areas. Environ Earth Sci 64(1):133–142

Santos DS, Mansur KL (2024) The educational importance of geosites representing geodiversity-biodiversity relationships: a thematic inventory proposal. Geoheritage 16(4):99

Selmi L, Coratza P, Gauci R, Soldati M (2019) Geoheritage as a tool for environmental management: a case study in northern malta (central mediterranean sea). Resources 8(4):168

Shrivastava MN, Sunil AS, Maurya AK, Aguilera F, Orrego S, Sunil PS, Cienfuegos R, Moreno M (2023) Tracking tsunami propagation and Island's collapse after the Hunga Tonga Hunga Ha'apai 2022 volcanic eruption from multi-space observations. Sci Rep 13(1):20109

Sischka L, Bosserelle C, Williams S, Ting JC, Paulik R, Whitworth M, Talia L, Viskovic P (2022) Reconstructing the 26 June 1917 Samoa Tsunami Disaster. Appl Sci 12(7):3389

Stojilković B, Gray M (2024) Geodiversification: the evolution of geodiversity through time. Geoheritage 16(3):91

Stojilkovic B, Stepisnik U, Ferk M, Valjavec MB, Cone S (2023) Geodiversity - foundation for life. Geogr Soli 31(2-3):23–37

Strandberg NA, Sear DA, Langdon PG, Cronin SJ, Langdon CT, Maloney AE, Bateman SL, Bishop T, Croudace IW, Leng MJ, Sachs JP, Prebble M, Gosling WD, Edwards M, Nogué S (2023) Island ecosystem responses to the Kuwae eruption and precipitation change over the last 1600 years, Efate, Vanuatu. Front Ecol Evol 11:1087577

Tukiainen H, Toivanen M, Maliniemi T (2023) Geodiversity and biodiversity. Geol Soc Spec Pub 530(1):31–47

Vegas J, Galindo I, Vázquez JT, León R, Sánchez N, Martín-González E, Romero C (2023) Tagoro, the youngest submarine volcano in the spanish geoheritage inventory: scientific value, geoconservation and opportunities for geotourism. In: Active volcanoes of the world. Springer, Cham, pp 257–272

Vernham G, Bailey J, Field R, Schrodt F (2024) What is the relationship between plant trait diversity and geodiversity? A plot-based, Pan-European analysis. Glob Ecol Biogeogr 33(11):e13904

Wei X, Zhang GL, Zhang J, Shi XF, Castillo PR, Zhang Y, Zhang WF, Xu YG, Li HY, Zhang H (2024) Overlapping hotspot tracks and melts from diffuse plume materials in the upper mantle generated intraplate seamount groups in the West Pacific. Earth Planet Sci Lett 643:118901

Witter JB, Self S (2007) The Kuwae (Vanuatu) eruption of AD 1452: potential magnitude and volatile release. Bull Volcanol 69(3):301–318

Woo KS, Kim L, Ji H, Jeon Y, Ryu CG, Wood C (2019) Geological heritage values of the Yongcheon Cave (Lava Tube Cave), Jeju Island. Korea Geoheritage 11(2):615–628

Yoon SH (2019) Geoheritage value and utilization plan of the Hanon volcanic crater, Jeju Island. J Geol Soc Korea 55(3):353–363

Yu Y, Yang J (2024) Advances in research on the impact of geodiversity on biodiversity. Adv Earth Science 39(4):347–356

Zakharovskyi V, Németh K (2021) Systematic literature review of the natural environment of the Coromandel Peninsula, New Zealand, from a conservation perspective. Conservation 1(4):270–284

Zakharovskyi V, Kósik S, Li B, Németh K (2023) Geosite determination based on geodiversity assessment utilizing the volcanic history of a near-sea-level explosive eruption-dominated volcanic island: Tūhua/Mayor Island. N Z Geol Soc 530(1):127–140

2 Geological and Geoheritage Setting of Samoa

The Samoan Islands are part of the broader SW Pacific region with their unique geological and geomorphological heritage sites dominated by their volcanic origin and their direct connection to the coastal and marine environment of the Pacific (Németh and Gravis 2022). The south-west (SW) Pacific is an expansive oceanic region with many of its islands difficult to access and relatively unexplored in a geoheritage context. Samoa is among the larger islands with both surface area and population perspective with an estimated human occupation span over 3 millennia (Addison and Matisoo-Smith 2010; Avis et al. 2007; Clark 1993; Clark and Michlovic 1996; Jackmond et al. 2022). This means that Samoa particularly rich in geocultural elements strongly linked to the natural environment (Sand et al. 2018). Many of ancient geocultural elements are still in practice of the everyday life and/or clearly traceable in traditions, customs legend. Many of such oral traditions are directly influenced or part of the broad aspect of geoheritage such as tattooing, dance, or various patterns in visual arts (Birkeland et al. 2008; Najita 2010; Németh et al. 2007; Schneider-Christians 1994). In a geotectonic context the SW Pacific is in the western extremity of the large. Old, and geological sense cold and rigid Pacific plate naturally segmenting apart through dominantly by various convergent plate margin processes such subduction and associated volcanic arc formation (Fig. 2.1). The plate margins are clearly visible along the entire Pacific marked by the location of earthquakes occurred in the past highlighting the zones where friction between the plates, magma generation and movement hence volcanism on the surface is more prominent (Fig. 2.2). The characteristics narrow zones where earthquakes and volcanism tend to occur along the Pacific Plates margin, commonly named as the Pacific Ring of Fire (Figs. 2.1 and 2.2). Alongside the dominantly subduction associated geotectonic elements, some large transform systems also act as displacing the edge of the Pacific Plate into plate fragments (Fig. 2.1). The reason behind this complex geotectonic situation is that the oldest part of the western Pacific Plate is far from its mid-ocean ridge spreading centre hence it had nearly 200 million of years to accumulate sediments to form thick (Boschman and van Hinsbergen 2016; Neall and Trewick 2008; Rowley 2008), cold and heavy edges along the old Pacific Plate region. Volcanism associated with plate boundaries is one of the most characteristic and recognisable geological processes linked with the Pacific in the public perception of the region. Volcanism is associated with arc-type geosystems forming complex and compound polygenetic volcanism in form of large stratovolcanoes emerging from the sea level offering majestic row of volcanic cones as islands where the subduction if prominent such as along Vanuatu or Tonga as part of the westward subducting Pacific Plate beneath the Indo-Australian Plate (Fig. 2.1). These stratovolcanoes are clearly majestic and due to their restricted surface areas, they often making them as focused geological heritage sites. While in modern, Holocene volcanoes across the SW Pacific arc systems their original volcanic landform is preserved, in older setting, significant erosion, coral reef formation and submergence created dissected volcanic islands commonly exposing deep internal part of former stratovolcanoes (Dickinson 1998; Ramalho et al. 2013; Thompson et al. 1998; Velmurugan 2008) such as the north-eastern part of Upolu Island and most of the American Samoan islands (Keating 1992; Stearns 1944a, b) (Fig. 2.3).

Beside the large stratovolcano systems and their eroded versions, other type of volcanism can also be recognized across the Pacific, most notable in Hawaii. Hawaii is a type locality of hot spot magmatism where deep sourced mafic magmas reaching the surface over long time, leaving behind hot spot tracks while the plate moves over the roughly stationary magma upwelling (hot spot) zones (Bachèlery and Villeneuve 2013; Clouard and Bonneville 2002; Dalrymple et al. 1977; Duncan 1990; Foulger 2007; Harpp et al. 2014; Hofmann and Farnetani 2013; Jackson et al. 2010, 2024; Koppers et al. 2011; Lipman et al. 2000; McDougall and Duncan 1980; McDougall et al. 1981; Price et al. 2014; Sinton and Sherrod 2020; Walker 1990) (Figs. 2.1 and 2.2).

K. Németh, V. Zakharovskyi, *Geodiversity of Samoa*, Geoheritage, Geoparks and Geotourism,
https://doi.org/10.1007/978-3-031-87220-4_2

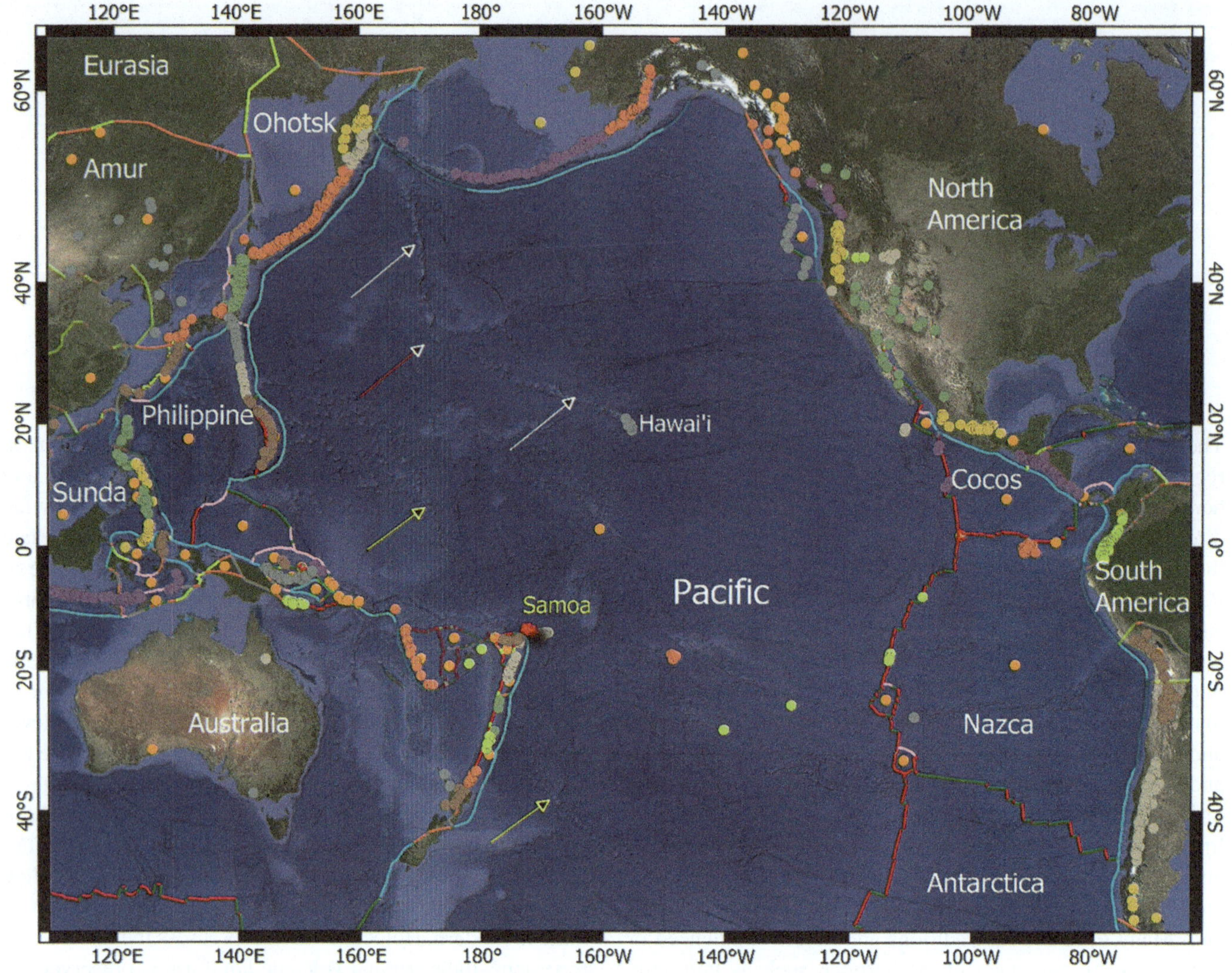

Fig. 2.1 Position of the Samoan Archipelago in the SW Pacific on a Google Earth Pro satellite image. Tectonic plate margins are shown by various coloured lines reflecting specific tectonic plate margin sections demonstrating the great diversity of the tectonic setting in the old SW Pacific region. The map showing with multicolour style Holocene volcanic groups defining the Ring of Fire circumference the Pacific Ocean basin. The white arrows point to the Hawaii hot spot track (in the northern segment) forming the Emperor Seamounts while in the south region the Northwest Hawaiian Ridge. Along that line the red arrow marks the kink of the hot spot track indicating the change of plate motion direction that moved over a stationary hot spot location that is currently beneath the Hawaiian island group. Samoa has a seemingly special geotectonic potion as being along a long hot spot trail (yellow arrows) while also nearby a sharp bend of the westward subducting Pacific Place deep trench and subduction zone. Geotectonic data is from Bird (2003). Holocene volcano groups are from Volcanoes of the World—Global Volcanism Program https://volcano.si.edu/ [accessed on 10 October 2024]

Such hot spot tracks have been recently recognized across the Pacific Plate and Samoa is inferred to be sitting over one such track, hence Samoa is geologically closely related to Hawaii (Clouard and Bonneville 2002; Price et al. 2014, 2022). This unique geological setting of the SW Pacific offers unique geological-geomorphological elements shaping the visible geodiversity of the region (Németh and Gravis 2022). However, what makes the SW Pacific truly unique is that all these geomorphological and geological heritage features are deeply linked to the living Polynesian culture, keeping traditions alive end deeply embedded to their natural environment (Fepuleai et al. 2017; Németh and Gravis 2022).

Both Savai'i Island and Upolu Island of Samoa have numerous potential geosites that could form the basis of any geopark project in various scales from local, regional or global (Fepuleai et al. 2021). In the last few years because of intensive research to identify Samoa's geoheritage values, the island of Savai'i was specifically selected during the Samoa Geopark Project Phase-1 as a location where geosite inventories has been developed as a first order approach to

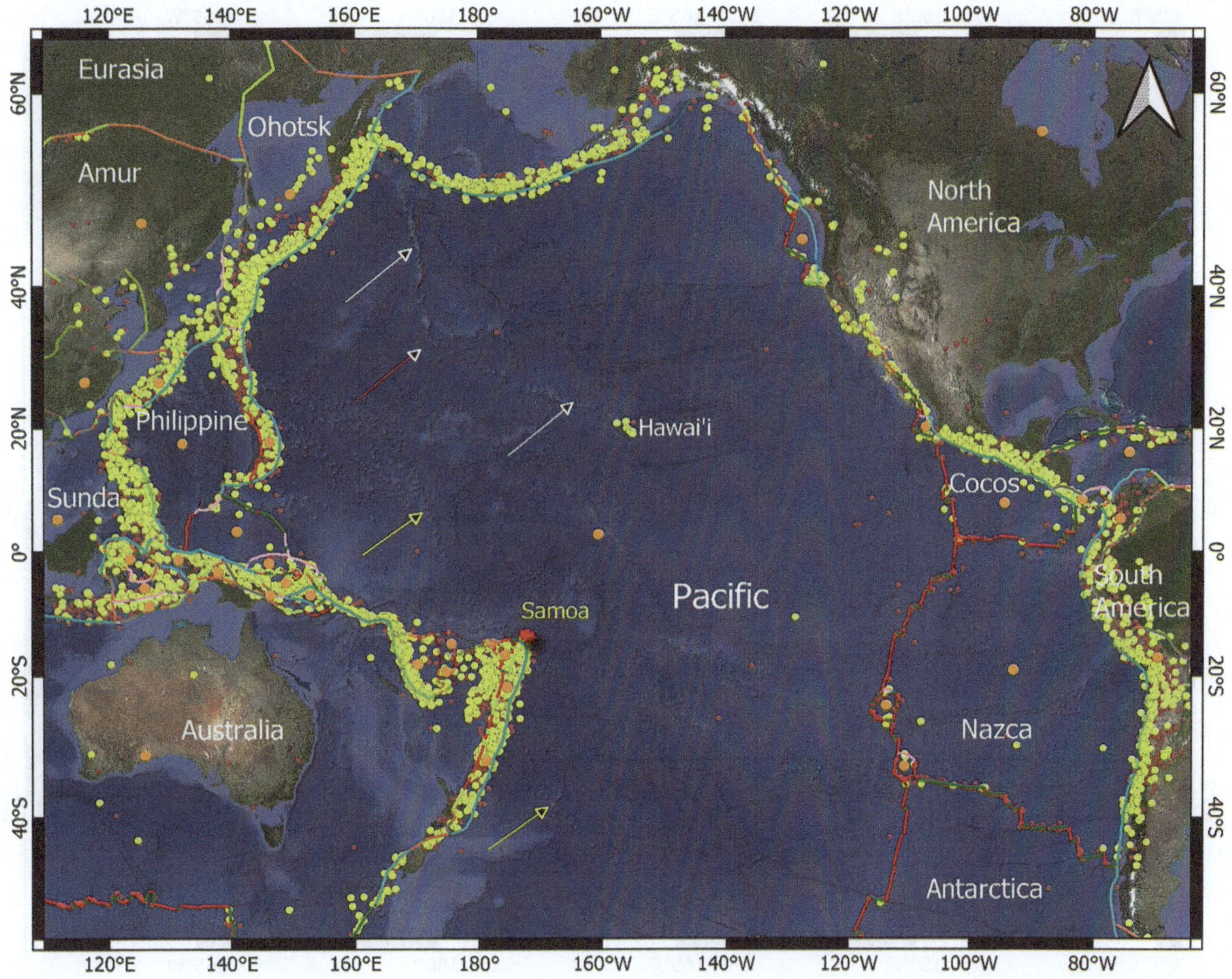

Fig. 2.2 Recorded earthquakes larger than 6.5 magnitude from the USGS Earthquake catalogue showing every earthquake recorded between 1900 and 2024. The earthquake pattern outline the tectonic plate margins putting Samoa in a sharp bend of the subducting Pacific Plate where typical large magnitude subduction earthquakes can form posing significant tsunami hazard for Samoa. Orange dots represent the center points of the identified tectonic plates. Major tectonic plates are named on the map

create a scientific basis for future geoheritage, geoconservation and geotourism ventures (Fepuleai et al. 2021; Németh et al. 2017). The recently initiated Samoa Geopark project's first focus was on the young large island of Samoa, where traces of Holocene volcanism is evident, and the natural habitat is intact in most places. Potential geosites were identified in a non-systematic way, mostly based on locating areas where local communities maintain legends and associated sites where current tourism operation is focused. So far, as in 2024, no systematic geosite inventory been created yet, despite the intensive research the island gained in the last decades mostly on developing geodiversity estimate methods that can be tested in island environments (Zakharovskyi and Németh 2021, 2022). The Samoan Islands are dominated by volcanic geoheritage elements from micro (e.g., specific lherzolite nodule locations) to island-wide scale (e.g., volcanic landforms). In addition, as tourism almost entirely linked to some beach activities, Samoa coastal geoheritage is also unique in sense of their diversity and abundance however it has never been inventoried following modern methods (Nunn 1998; Richmond 1992; Zann 1994).

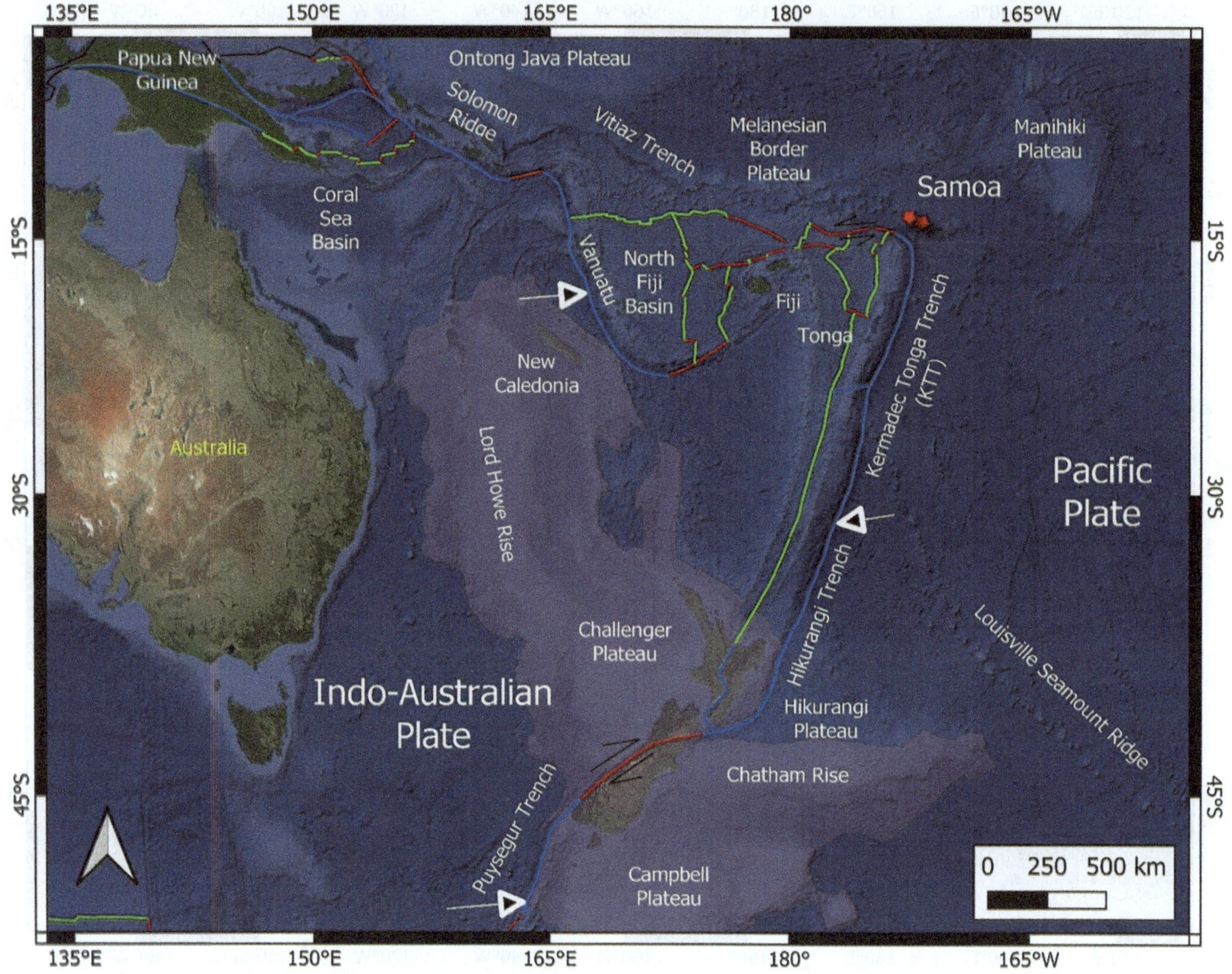

Fig. 2.3 Geotectonic situation of the proximity of Samoa in the SW Pacific geodynamic context. Tectonic plate boundaries with their type of colour coded. Blue lines refer to convergent plate margins (e.g., subduction), red lines indicate transform systems (e.g., strike slip movement) and green lines indicates divergent plate margins (e.g., extensional basins such as back arc basins). Main plate movement directions are marked by arrows. The mostly submerged continental lithosphere of Zealandia (its emerged part is New Zealand), a submerged microcontinent marked by a light purple region

2.1 Origin of the Samoa Islands

The Samoan Island chain is in the southern part of the Pacific Ocean to the northeast of New Zealand between latitudes 13° and 15°S and longitudes 186° and 191°W parts, Western (Independent or in early twentieth century German Samoa) Samoa and American Samoa surrounded by deep ocean (Davies et al. 2004) (Figs. 2.1 and 2.2). The actual geotectonic situation around Samoa is rather complex as the Samoan Island chain located to the northeast of the sharp bend part of the Kermadec Tonga Trench (KTT) (named as the Northern Terminus (NT)) (Fig. 2.3). The plate margin variation around Samoa is also great as all the types exists within relatively short distances (Fig. 2.3). The Pacific Plate is subducting westward beneath the Indo-Australian Plate forming well developed volcanic arc and associated back arc basins parallel with the Kermadec—Tonga Trench (Fig. 2.3). Local opposite intra oceanic subduction occur west from the Kermadec—Tonga Trench forming the Vanuatu Arc with their chain of large stratovolcanoes (Fig. 2.3). In between large oceanic basins such as the Fiji Basin occurs with complex crustal structures and with some older rocks exposed in the surface forming the base of the main islands of the Fiji archipelago. Due to the rapid orientation change of the Pacific Plate boundary around Samoa results transform plate boundaries to form. In addition, older, inactive trench and volcanic arc such as the Vitiaz Trench (Fryer 1974; Pearce et al. 2007; Wyn

Hughes 1978) (Fig. 2.3) indicates the location of former subduction fronts. In the broader geotectonic situation, the region south of Samoa, within New Zealand landmass is sitting is Zealandia Microcontinent (Mortimer 2020) (Fig. 2.3). Zealandia microcontinent is largely submerged and about 5% of its total surface area is above the current sea level (mostly New Zealand) (Mortimer 2020; Mortimer et al. 2017; Mortimer and Scott 2020) (Fig. 2.3). Its crustal structure is different than those upon Samoa located being more continental type with old crystalline basement rocks derived from Gondwana, when Zealandia separated in a series of rifting events from the landmass subsequently formed the Australian continent (Fig. 2.3). Zealandia is geologically a continent sitting over two tectonic plates separated by a long transform fault system (Alpine Fault) (Fig. 2.3) with a right lateral relative plate motion between Indo-Australia and the Pacific Plate (Fig. 2.3). Seafloor mapping and sample collection also showed that around Samoa three major oceanic basalt plateau located such as the Manihiki in the north-east of Samoa (Petterson 2020; Timm et al. 2011), the Hikurangi just east of the North Island of New Zealand (Mortimer and Parkinson 1996) and the Ontong-Java in the west of Samoa (Michael et al. 2024; Tejada et al. 1996; van de Lagemaat et al. 2023) (Fig. 2.3). These large basaltic lava plateaus form a shallow oceanic region clearly visible on the bathymetry and covering areas comparable to the size of the landmass of the main islands of New Zealand (Fig. 2.3). Their origin is the direct result of massive and rapid (in geological sense) outpouring of basaltic magma to the sea floor due to mantle plume activity that generated large igneous or flood basalt provinces like those known in India, South Africa or South America, with a difference that these ones formed in subaqueous conditions (Fig. 2.3). The significance of these large basalt plateaus on the oceanic crust that due to their significant thickness and cold, rigid nature they can't easily subduct. As a result, for instance the Hikurangi subduct in low angle and locking into the continental lithosphere holding the North Island of New Zealand as part of Zealandia (Okuma et al. 2022; Shreedharan et al. 2022; Wang et al. 2023). Similarly, these large basalt plateaus can act as an "insulator" making difficult to pierce magma through if such zones crossing a mantle hot spot (Jackson et al. 2024).

Samoa itself sitting on a former hot spot track (Jackson et al. 2024). In the sea floor, other hot spot tracks are clearly visible such as the hot spot track Hawaii sitting on its terminal point (Figs. 2.2 and 2.3). South of Samoa other hot spot track can be traced that forming submarine volcanic islands, guyots, such as the Louisville Seamount Ridge (Fig. 2.3).

The current tectonic situation around Samoa showing clearly the zones where the tectonic stress is compressional and zones where due to the subduction roll back local extension can form various sub-basins (Fig. 2.4). The current stress field pattern of the region also provides good explanation, that Samoa is in the tip of a major edge or corner of a tectonic plate where sudden changes of the tectonic stress field orientation is expected (Fig. 2.4). Samoa is in a close vicinity of this tip that can be viewed as a major geohazard to the Samoa Island in form of large subduction earthquakes to form that may be able to trigger large tsunamis (Riou et al. 2024; Sischka et al. 2022; Williams et al. 2021, 2022) (Fig. 2.4). Tsunami hazard in Samoa, especially in its southward facing coastal area is significant as that coastal area has limited reef coverage (Williams et al. 2021). Especially in its south-eastern coast of Upolu has just a narrow shallow coastal area hence tsunami not only can reach very quickly the coastal regions but also can be very energetic as well as the 29th September 2009 tsunami demonstrated (Couling 2014; Franchello and Annunziato 2012; Gokon et al. 2014; Tonini et al. 2011; Williams et al. 2021). Tsunami destructions are deeply impacted local coastal communities (Seiuli et al. 2016) and hazard education is an important aspect where geoheritage can be incorporated.

The Samoan Island chain has 1400 km in length with a width of up to 380 km. It aligned northwest to southeast along the north-east of the sharp bend of the Kermadec-Tonga Trench (KTT) region, known as Northern Terminus (Fig. 2.3). The KTT represents a boundary between the two major plates, Indo-Australian Plate and Pacific Plate where the Indo-Australian Plate currently moves north-east at 6.6 cm/year speed while the Pacific Plate propagate westward at 7.1 cm/year speed (Hart et al. 2004; Hawkins 1987).

Immediately to the south of Samoa, the northern end of the KTT bends sharply to the west, to become part of the east-west aligned Vitiaz Lineament, to the north of the Fiji group (Fig. 2.3). The Vitiaz Lineament formed as the Indo-Australian Plate continues to advance North. Its margin drags the tearing section of the Pacific Plate in a mechanism defined as down-drag subduction (Hawkins 1987; Natland 1980; Natland and Turner 1985). In other words, the Vitiaz Lineament represents the torn lithospheric part of the Pacific Plate (Hart et al. 2004; Hawkins 1987). The highly oblique nature of the Vitiaz Lineament generates a range of transpressional phenomena from Papua New Guinea through the Solomon Islands to Samoa (Fig. 2.3) (Petterson et al. 1997, 1999). This tectonic architecture somehow diminishing toward the modern Samoa Island chain, however, on the sea floor just west of Samoa still recognizable such geometry pattern.

Fig. 2.4 Current tectonic stress regime from the World Stress Map Project (https://www.world-stress-map.org/) around the Samoan Island chain

2.2 Origin of Samoan Volcanism

The interaction of the two major tectonic plates nearby the Samoan Island chain is the main source of earthquakes occurred in the region (Hawkins 1987; Natland 1980; Natland and Turner 1985). These earthquakes are typical subduction earthquakes with great tsunami-generating potential. Among these earthquakes that are larger than magnitude 6 are not uncommon, and even super strong motion has also been recorded in the recent history. These include the two recent earthquakes with magnitudes of 8.0 and 7.9 which generated the 29 September 2009 tsunami that killed 189 people (Franchello and Annunziato 2012). This tragic event has triggered new research not only on looking for traces and evidence of ancient tsunami events in Samoa but also looking at other mass movement processes influenced the landscape evolution of the Samoan islands (González-Riancho et al. 2015; Richmond et al. 2011; Tonini et al. 2011; Williams et al. 2021).

As the Samoan Islands are largely volcanic in origin, the tectonic activity and their potential implication to trigger volcanism is a major concern of the natural hazard landscape of Samoa (Cronin et al. 2000, 2006; Németh and Cronin 2009a, b). Beside the hazard studies that are not only restricted on the elevated nature of volcanic hazards, but evidence also show high susceptibility of landslides, rock falls (Fepuleai and Németh 2019; Goodwin and Grossman 2003) or flooding in the narrow volcanically determined gorges and canyons to the coastal plains (Wu et al. 2021) are among the key elements of Samoa hazard outlook. These geological events and their deposits and landscape together form a rich geoheritage assets of Samoa that is currently underutilized and underappreciated. The island volcanic geoheritage is significant as all the geological forms associated with oceanic lava shield volcanism with all their stages are well exposed and well represented, most of them accessible. However, heavy vegetation often prevents easy visitation to the main dorsal ridges. There are exceptions where scoria cone complexes, many of them with some crater lakes, are clearly visible in their well-preserved original landforms of scoria cones such as Mount Fiamoe, just offside from the Cross Island main road of Upolu (Figs. 2.5 and 2.6).

The Geological map of Western Samoa distinguished six formal lithostratigraphy formations associated with major and minor faults on the two main islands based on the Kear and Wood (1959) geological map (Kear and Wood 1959). Both islands are crowned by typical small volume volcanic cones (e.g., scoria cones) (Figs. 2.5 and 2.6) in the axis of both Savai'i and Upolu Islands (Fig. 2.6). Along the main dorsal ridges bounding fissures are identified mostly based

Fig. 2.5 Typical scoria cone complex, Mount Fiamoe (13°55′24.16″S, 171°47′55.76″W) in the top of the dorsal ridge of Upolu Island of on a Google Earth Pro Satellite Image looking toward NW

on satellite imagery as the ground in most cases heavily vegetated. These zones are informally named to Savai'i Major Fissure System (SMFS) and Upolu Major Fissure System (UMFS) indicating a fissure origin of the volcanic chains of both islands (Fig. 2.7). The two fissure systems are intuitively connected through the Inter-Island Fissure System (IFS) that is parallel with the line along Manono and Apolima Islands located between the two major islands (Fig. 2.7). Again, this fissure relationship is very theoretical and only based on the vent distribution and the limited bathymetry data available (Fig. 2.7).

The volcanic evolution of the Samoan Island is largely attributed to the typical lava shield and atoll formation theory based on the model applied for Hawaii in the Pacific region (https://www.usgs.gov/observatories/hvo/evolution-hawaiian-volcanoes). In this model eight stages can be distinguished as a combination of various type of volcanic phases culminated to an erosion and submergence of the volcanic islands (Schmincke 2004). This model commonly named as an ocean island evolution stages consists of: (1) initial—deep submarine stage, (2) submarine—shield building stage, (3) near sea level—shield building stage, (4) subaerial—shield building stage, (5) capping stage, (6) erosional stage—reef growth, (7) renewed or rejuvenated volcanism stage, and (8) atoll stage. Since the first time this concept was introduced based on the geomorphological research of the Hawaiian island volcanoes by Harold Stearns (Stearns 1946), the same model applied for most of the ocean island basalt volcanoes, not only in the Pacific but everywhere (Esson et al. 1970; Kwasnitschka et al. 2024; Smith et al. 1989). Ocean island basaltic volcanoes evolve through three major volcanological stages with distinct geochemical and petrological signatures defined as lava shield, postshield, and rejuvenated volcanism. Of course, this classification seems easy, it is far more complex to be able to recognize and distinguish such stages as the eruptive products and their physical appearance (e.g., style of volcanism) are commonly very similar to each other. Moreover, in case of overlapping ocean island volcanoes, complex geological architecture can develop such as the case in Hawaii for instance. The identified stages can be repeated laterally, making the geological mapping even more complicated such as the case in the Samoan islands. For the original triplicate distinction another stage of volcanism commonly referred to such as a fourth stage, the preshield stage. The volcanic stages are somehow associated with the erosional stage, lowering, partial submergence and a complete atoll stage where coral reef development on the remaining volcanic edifice became an important

Fig. 2.6 Mount Fiamoe in Central Upolu's dorsal range is a typical scoria cone complex dominates the landscape over Apia. The view is looking toward south from the point of 13°55′2.02″S, 171°47′22.32″W. Note the exposed basaltic blocky lava surface as an old morphological level mapped as part of the Middle Pleistocene Salani Volcanics, while the Mount Fiamoe is one of its same age volcanic cones

phase (Petterson 2020). The geomorphological stages and evolution while associated with the volcanic history, somehow different from what the geological stages refer to. The geological evolution is somehow linked to the time needed a hot spot that provide mantle upwelling gains enough momentum and energy to be able to pierce the relatively thin oceanic crusts and feed melt to the sea floor in a start of the shield stage. This process can take place over 100s ka time and can be sporadic submarine volcanism (preshield) prior the steady and continuous upwelling of melt that can feed substantial volume of melt to the seafloor to build a lava shield that can reach a subaerial stage and form an oceanic island. In this model the postshield stage refers to the time when the already grown volcanic island melt production reaching its depletion due to the plate movement over the (near) stationary hot spot, gradually forcing the melt to be diverted from the long-established conduit system or due to the fluctuation of the melt production within the same mantle upwelling that can have 100s ka time variations. As a result, the post shield volcanism form relatively small volume volcanic landforms, gradually filling an increasingly subdued landscape. In the absence of volcanism, the surficial processes take over the landscape evolution and generate typical and marked erosional horizons and common coral reef formation along the lowered volcanic islands.

Interestingly, it has been observed in numerous occasions that large oceanic lava shield volcanoes are commonly associated with some sort of volcanic cap that are not only separated by 100s ka time from most of the volcanic evolution of the shield but also different in its geochemical signature (Garcia et al. 2016, 2024). Due to the geochemical variations that can and used to be applied to identify specific stages of a the lithostratigraphy of a lava shield complex, and using those markers as locations where significant erosion stage operated in the evolution of the oceanic island (Garcia et al. 2024).

The rejuvenating stage of volcanism in the life of oceanic island volcanoes is an enigmatic question and subject of heated geological debates. The main question is, why a volcanic island that has been active over 100s ka time and produced 10s of km^3 lava mostly in the form of extensive mafic (e.g., basaltic) lava flows.

Fig. 2.7 Fagaloa Bay on a Google Earth Pro satellite image in the northern sector of Upolu resembling a volcano collapse scar carved into the oldest exposed volcanic rocks of Samoa. Some coarse grained, thickly bedded breccia exposed in scattered outcrop that could be volcanic debris avalanche deposits, or volcaniclastic assemblages associated with volcanic mass wasting, but so far, no compiling and convincing evidence to be sure of such interpretation

Of the four stages, the hardest to explain is rejuvenation, in which volcanoes resume their activity 1–2 million years after the postshield stage ceases. As in the rejuvenated stage much smaller volume of magma reach the surface the resulting volcanic landforms as scoria cones, tuff ring and tuff cones are more distinct features and can provide additional element of the geoheritage of an oceanic island volcano.

There are three major explanation exists for the origin of rejuvenated volcanism, but it is likely that these three main tracks of explanations can work together. One of the simple explanations dealing with the overall weight of the growing oceanic lava shield that sitting over a thin oceanic. The weight of the established lava shield sink into the mantle while in its marginal region as a counteraction, mantle can slowly fill the place. The mantle protrusion into shallower level, hence lower pressure can initiate small volume of melting. This model explains well why 100s ka to few millions of years gap exist between the main phase of lava shield growth and the rejuvenation, as such amount of time is expected to generate such low volume of partial melting of the mantle. However, in most cases the time gap between shield and rejuvenated stage is significant, there are examples where this transition took place rapidly (Lara et al. 2018) making this question even more enigmatic.

It is important to consider in any explanation of rejuvenated volcanism on oceanic volcanic islands is that the plate motion slightly shifting the relative position of the main heat source which is theoretically enough to make some small part of the mantle go through melting due to the locally developed undulating base of the crust. Alternative theory based on some research on Hawaii suggested that the rejuvenated volcanism is rather the result of the release of some melt residue from those melts that never been able to contribute to the shield phase. This model therefore offering a solution that is in a complete opposite way to the previous models as the gravitational bending-triggered partial melting operates by old magma residue and its release.

Model predictions also suggested that the geometry of a mantle plume that is heating and providing rising melt in its centre toward the surface over a slow mowing plate also have some convective downward movement of material in slow speed. The plate movement can gradually shear off melt pockets from the centre of the plume allowing it to drift aside potentially 100s of kilometres away from the centre point of the rising heat source. Some numerical model that include the combination of upward moving hot material and melt generation with the drag effect and downward push of the side of the hot material demonstrated a physical possibility to have conditions long after the plate moved over a hot spot to have some melt pockets to successfully reach the surface and form more like a monogenetic type of (short lived, low volume and sporadic) volcanism (Németh and Kereszturi 2015) we observing as rejuvenated volcanism. This model is essentially operating with the concept of the localized melting events.

Rejuvenated volcanism on oceanic volcanic islands has also been associated with sea-level changes that affecting the overall load on the magmatic plumbing system likely able to produce small volume of (monogenetic style) melting (Jicha et al. 2023).

The interaction of the two major plates at the sharp angle causes a portion of the Pacific Plate to be torn, suggesting that it generates rejuvenate volcanism in Samoa (Natland 1980). This tearing part of the Pacific Plate forms a down-drag sub-duction, as the Indo-Australian Plate advance toward north-east (Fig. 2.3) (Hart et al. 2004; Hawkins 1987). Hart et al. (2004) uses the term "rejuvenated" to describe post-erosional activities dominating the later (young) subaerial and submarine volcanic cone. Rejuvenated eruption is the stage where a volcano essentially refuels or recharge its melt storage network after a long period of dormancy and its consequent erosion. It is suggested that Holocene volcanic activities of the post-erosional stage spread along the island chain of Samoa (Kear and Wood 1959; Natland 1980; Natland and Turner 1985). It is a long-established model to look at Samoan volcanism along with some lineal progression that is associated with a stationary hot spot activity and the moving plate. The evolution of the Samoan volcanism commonly suggested to be very similar to that of Hawaii (Stearns 1944a, b). While in Hawaii however, the specific stages of the pre, shield and rejuvenated volcanism are consequently showing a systematic order (even by factoring in the potential overlapping multi-shield aspects) (Stearns 1946), the case in Samoa is way more complicated as there is no clearcut age or spatial or even stratigraphical progression (Johnson et al. 1986; Koppers et al. 2008; Natland 1980). In a large scale it seems that there is some sort of shield development, but it is way not as clear than in Hawaii, despite the lava shield morphology is somehow evident in the case of Savai'i, but very obscure in the case of Upolu or in American Samoa. The age progression discrepancy is commonly explained by the impact of the nearby Tonga—Kermadec Trench fissures can open complicating the magmatic system fed by complex plume upwelling (Konrad et al. 2024; Koppers et al. 2011).

Samoa's oldest volcanism can be divided as Pleistocene to Pliocene and part of the Fagaloa Formation (Table 2.1). The age difference often manifested in a subdivision of Lower Fagaloa and Upper Fagaloa sub-formations as a reflection of the pre-shield and shield stage of the volcanic evolution as Lower and Upper Fagaloa Formation respectively. The widely accepted concept of Kear and Wood (1959) implies that major erosion occurred as the major shield activities ceased (e.g., after the Fagaloa Formation deposition) and manifested in the development of a major unconformity or erosion surface believed to be traceable across the two islands. The post-erosional volcanism then

Table 2.1 Lithostratigraphy of Upolu and Savai'i

Lithostratigraphy—formation	Age	Lithology
Aopo volcanics	Historical	Cones erupted in last 200 years, fresh porphyritic pahoehoe flows and a'a flows common only around cones which fill older valleys and spill out over coasts to fill lagoons and cover the barrier reef
Puapua volcanics	Middle to Late Holocene	Basaltic lavas (pahoehoe) and well-preserved cinder cones with scoria and scoriaceous basalt Thin soils: lavas flow offshore and form rocky (or "ironbound") coasts; ubiquitous aa and pahoehoe structures form broad domes
Lefaga volcanics	Early Holocene	Feldspathic porphyritic basalts (picrite basalts and dolerites) and thick irregular scoria beds (a'a) with many bombs and lapilli Lack of dissection; only narrow fringing reefs present offshore of these outcrops
Mulifanua volcanics	Last Glacial (Late Pleistocene)	Vitreous, porphyritic, and non-porphyritic basalts (olivine basalts, dolerites, and analcite basalts) interbedded with a'a. The upper flow surfaces typically consist of pahoehoe The existence of wide barrier reefs existing offshore of these outcrops; only shallow stream channels
Salani volcanics	Last Glacial and older (Late Pleistocene?)	Picrite basalts, and olivine chlorite and basalts that typically grade upward from porphyritic basalt through vesicular basalt to rubbly a'a Deep soil and weathering Evidence of pre-Mulifanua canyon cutting
Vini Tuff	Late Pleistocene (?)	Hard, thick-bedded, calcareous tuff (vitric)
Fagaloa volcanics	Late Pliocene or Early Pleistocene	Steep weathered slopes of intercalated a'a and pahoehoe flow rocks, rubbly scoria, ash beds, vitric tuffs, and contemporaneous basaltic dykes

rather rift-aligned and generated a characteristic morphostructural elements to both islands and provided the mainframe of their roughly NWN-SES trending pattern of main rock units. Along this rift zones there is no clear spatial or temporal pattern of occurrence of volcanism, they seem to appear randomly in both islands. Such rift related distribution of volcanism is not unknown in other islands in the SW Pacific, especially along the Vanuatu Arc's Ambrym and Ambae volcanoes (McCall et al. 1970; Németh and Cronin 2009a, b) where along a central shield, rift-aligned young cones can be traced over tens of kilometres length aligned with the main morphostructural trends known from the seafloor. Along this rift zone dry explosive small volume volcanism are typical in the elevated areas such as scoria and spatter cones, while in lowlands tuff ring, maars and tuff cones are common, representing the whole range of explosive hydromagmatism and their typical eruption styles. This conceptual framework of volcano types in the post shield stage Samoan volcanism is apparent and share similarities to common rift aligned monogenetic volcanism type without even having characteristic lava shield associated with it such as in Taveuni in Fiji (Cronin et al. 2001).

Large oceanic shield volcanoes in deep marine basins are commonly associated with major mass wasting stages when entire sector of the lava shield collapsed and fed submarine landslides and complex sedimentary systems. Submarine debris avalanche on the south, north, and south-west of Samoa inferred to correspond with similar landslides documented from Hawaii or most of the Canary Islands in the North Atlantic (Boulesteix et al. 2013; Goodwin and Grossman 2003; Keating et al. 2000). however, there is no clear mappable avalanche deposits preserved or known from neither Savai'i nor Upolu. In Upolu, at Fagaloa Bay (from where the Fagaloa Formation names come from), the oldest volcanic rocks are exposed in Samoa, and they are commonly thick, large volume volcanic breccias (Fig. 2.7). These rocks have not been mapped yet in detail, nor established their precise origin. They are likely to be the proximal part of a larger composite volcanic edifice that went through significant erosion since the Pleistocene. In scattered, mostly coastal exposures large angular breccias are resembling mass wasting processes of their origin that could be the remaining proximal part of volcanic debris avalanches as some mesotextural elements (e.g., jig-saw fit texture) are indicative for such origin. However, they also could just be proximal crater or vent filling chaotic breccias typical in association with large volume polygenetic volcanoes.

More prominent feature is in the island of Savaii's northern sector. A massive scar-like feature with curving morphological step closely resembling a volcanic debris avalanche amphitheatre that removed the northern sector of the main lava shield (Cronin et al. 2000, 2006) (Figs. 2.8 and 2.9). This is a feasible assumption but there is almost no direct evidence to support it as there is no extensive volcanic debris avalanche deposits ever been mapped in the region. However, the absence of evidence is not obviously the proof of the lack of such rock units. Volcanic debris avalanche can be removed by surficial processes rather quickly, leaving an open amphitheatre behind or, most of the deposits moved well beyond the current landmass of Savai'i and the deposits are in the seafloor several tens of kilometres offshore of the northern coast of Savai'i. Unfortunately, there is no convincing bathymetry or seafloor dredging data to support this theory, hence it remains a rather speculative idea now.

Fepuleai (2016) suggested that major and minor faults on the two major islands were part of a cone-collapse event (CCE) inferred to be occurred sometime at least 22.3 ka (Fepuleai 2016). While this is a rather speculative estimates, there are some evidence indicatives for volcanic instability processes to be occurred in Samoa, however their scale is hugely variable from cone-scale to island scale. Indeed, there are plenty of evidence for mass movements in the volcanic terrains not only in Samoa, also on the islands of American Samoa as well. This collapse events however difficult to assign to specific stage of the island evolution and many of them are rather local cone collapse features, as part of the nature how such scoria cones for instance could evolve if they erupted on a steep terrain, hence the significance of their cone collapse identification is unclear at this stage.

There are cone-collapse features such as (1) collapsed scoria cones of Fagaloa Formation of Upolu and Salani Formation of Savai'i, (2) upthrown (up to 60 m) of the lavas of the Salani Formation in the easternmost part of Upolu and northern portion of Savai'i, (3) collapse-like feature often named as the Fagaloa Depression that informally referred commonly to as a caldera in northeast Upolu (however no evidence has been showed to support caldera origin of the depression from the geological record), (4) volcanic flank collapse of Tau Island (American Samoa), (5) oceanward flank collapse on off-shore volcanic islands (tuff cones) in the easternmost part of Upolu, and (6) a some submarine volcaniclastic deposits recognized on bathymetry survey inferred to be volcanic debris avalanche deposit in the north eastern part of Savai'i.

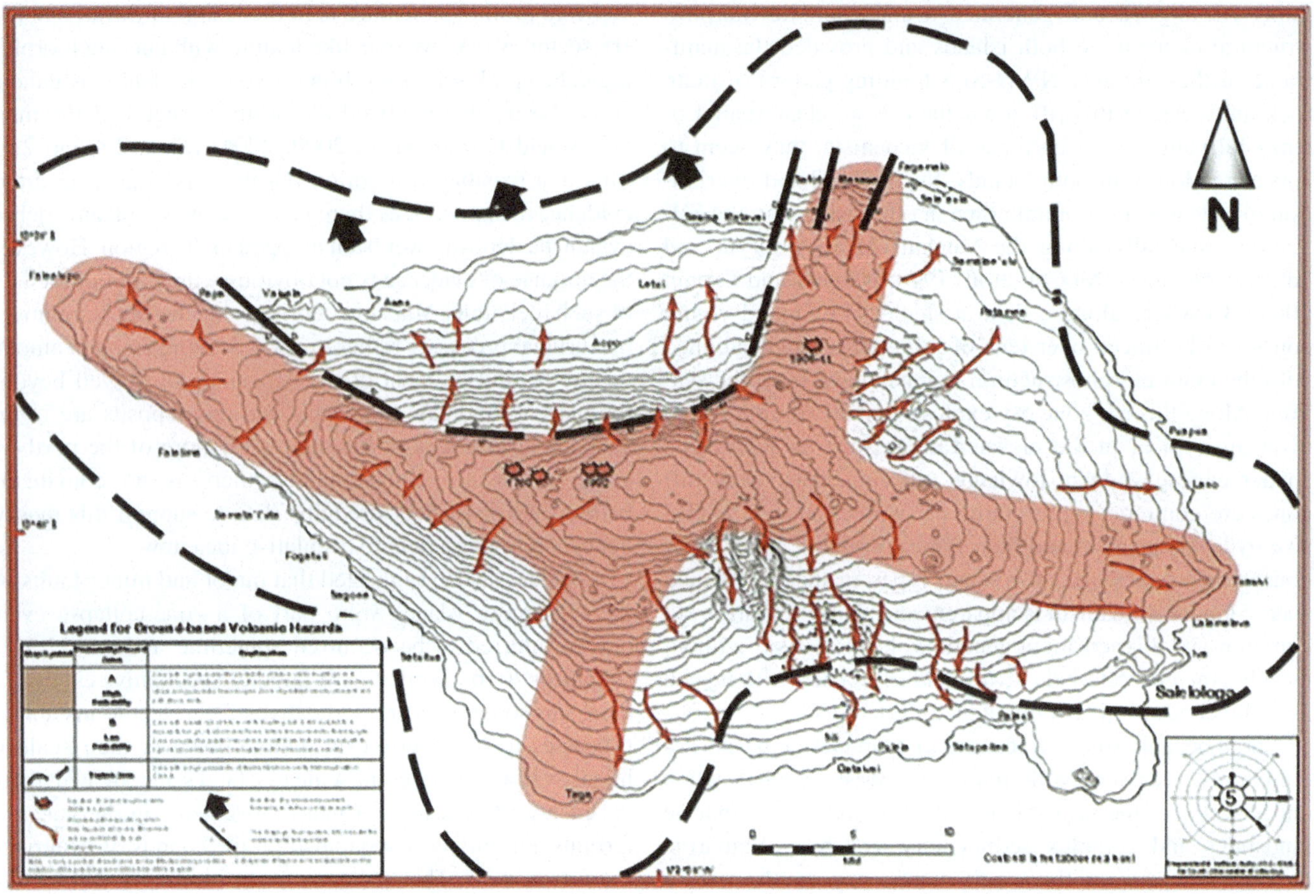

Fig. 2.8 Geohazard map of Savai'i based on the location of volcanic vents and major structural elements identified on the satellite imagery (Cronin et al. 2000). Note the significant curved area in the northern region of Savai'i closely resembling a volcanic collapse scar

Fig. 2.9 3D view on a Google Earth Pro satellite image toward the northern scar feature in Savai'i that is distinctly featuring some mass movement such as volcano collapse. The scar amphitheatre collected the youngest lava flow fields in the region. Yellow line marked the main road of Savai'i roughly following the curved topography

2.3 Main Geological Units

Volcanism in Samoa is traditionally subdivided from stratigraphy point of view into six formal lithostratigraphic units named as unique geological formations (Kear and Wood 1959) from the youngest to the oldest: Aopo Formation (1760–1911), Puapua Formation (Middle to Late Holocene), Lefaga Formation (Early Holocene), Mulifanua Formation (Late Pleistocene), Salani Formation (Middle Pleistocene), and Fagaloa Formation (Middle Pleistocene to Pliocene) (Table 2.1 and Fig. 2.10).

The Fagaloa Formation collect the oldest rocks known from Samoa that represents the main stages of the oceanic lava shield growth period in the Pliocene-Pleistocene, well before subsequent post-shield volcanism started. Following the Fagaloa shield-building phase, a volcanic quiescence took place that allowed tropical erosion to take place and generate significant erosional surface and lowering the original shield to generate a subdued landscape (Table 2.1).

Following the major erosional stage, post-erosional volcanism took place in well distinguished periods leaving behind volcanic rocks formally mapped as the Salani Formation, Mulifanua Formation, Lefaga Formation, Puapua Formation, and Aopo Formation (Fig. 2.10 and Table 2.1). Stratigraphically distinct rock formation of various pyroclastic successions typically associated with various type of magma and water explosive interaction leading the formation of tuff cones along the rift edge of Uplolu (both in their eastern and western edge) formed volcanic islands that are heavily modified by wave cut erosion. These volcanic landforms are clearly not part of the major lava shield due to their very different eruption mechanism responsible for their formation, and their individual volcanic edifices, their erosion stage indicates that they rather part of the late stage of the shield-building eruptive history, hence stratigraphically they are placed between the shield stage Fagaloa Formation and the youngest post-shield volcanics. While this is a logical compromise to assign these rocks to specific stratigraphy units, there is very little direct evidence (e.g., direct age dating) that can explicitly proof their stratigraphy position hence their role in the volcanic edifice growth of Upolu (Fig. 2.11).

The island chain of Samoa has been dominated by alkalic volcanic activities (Cibik 1999; Fepuleai 2016). The geochemical nature of lavas from deep submarine and subaerial volcanoes, of the Samoan island chain, reveals a combination of shield and post-erosional volcanism without clearly

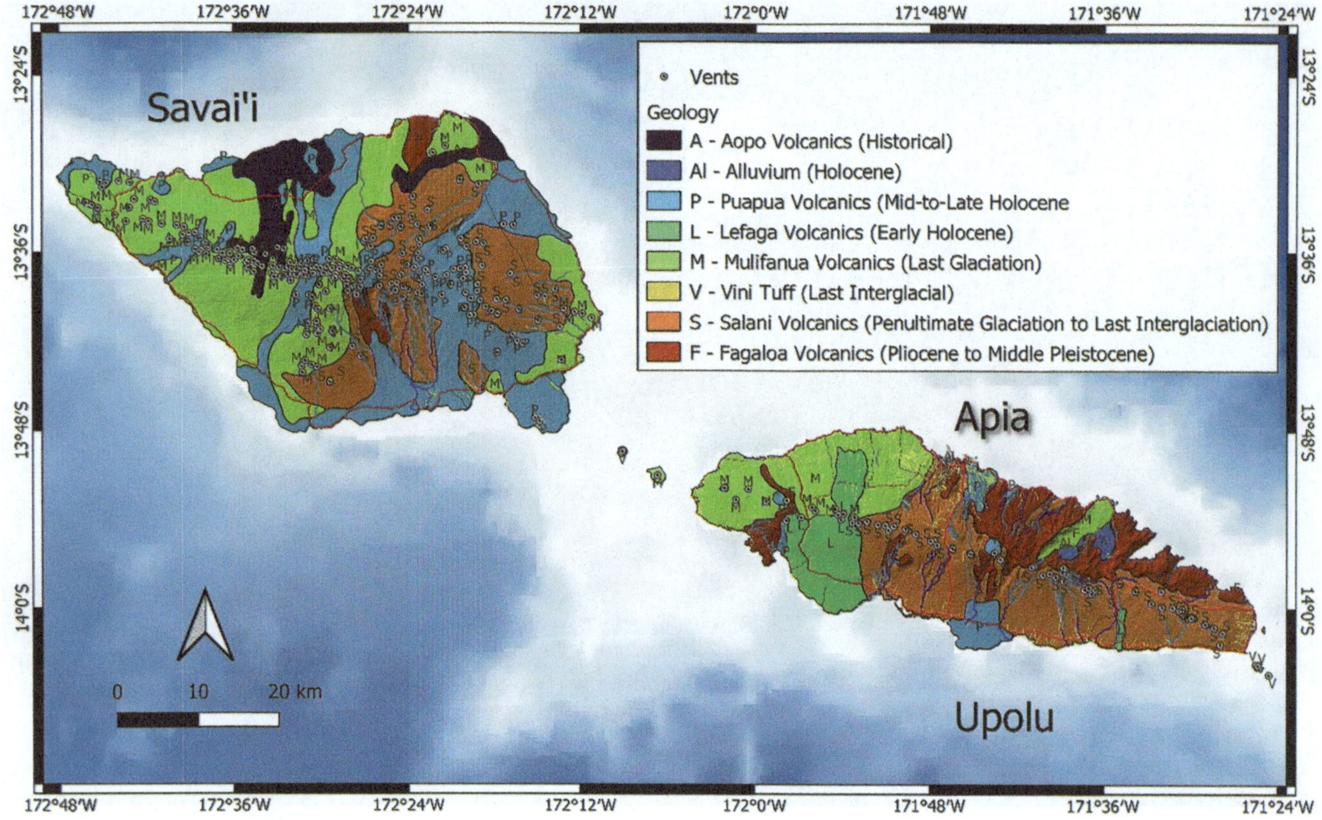

Fig. 2.10 Geology map of Samoa (Kear and Wood 1959)

mappable evidence of spatial-temporal zones (Fepuleai 2016). Upolu and Savai'i are deeply dissected by series of major and minor faults such as the Sataua-Ologogo Arc Fault (SOAF), Safune Fault (SF), and Manase-Gataivai Fault (MGF) seem cross cutting each other to the northern part of Savai'i. Vaisigano Fault (VF), Fagaloa-Falealili Fault (FFF), and Fanuatapu Fault (FF) are more like elongated northeast southwest. Please note that these fault locations are based on aerial photos and satellite imagery and there is very little evidence to explicitly identify their surface expression or map them on the ground. This issue is largely due to the heavy vegetation and the complex lava coverage of the island.

There is evidence of subsequent dyke intrusions in some places such as the area of the exposed rocks (mostly along coastal cliffs) of the Fagaloa Formation. In places a series of basaltic dikes are prominent features and informally named as the Fagaloa Intrusion as a succession of cm-to-dm wide dyke swarms (Fepuleai 2016). The dyke swarm however is rather a local feature, and it is difficult to capture their significance as similar diking is the nature of lava shield volcanism and expected to be everywhere. Due to outcrop availability, we are just not able to see them in other places. Typical chilled margins of many of the exposed dykes suggest time difference between the intrusion and the host volcanics to allow the melt to get solidified rapidly. A massive dike system called Lemafa Intrusion believed to be extends from Falevao along Fagaloa Bay and east toward the Uafato and Ti'avea village (Fig. 1.3) forming a characteristic zone of about 15–100 m wide and traceable over 10 km length.

The Lemafa Intrusion as a part of a deeply weathered intrusive body associated with the strongly eroded shield volcano manifested in volcanic rocks grouped in the Fagaloa

Fig. 2.11 Nu'utele Island (14°3′48.25″S, 171°25′21.72″W), just offshore from eastern Upolu is a deeply eroded tuff cone dominated by typical hydromagmatic pyroclastic rocks formed due to Surtseyan style explosive eruptions. The tuff cone original morphology has been modified due to wave cut erosion, especially those areas facing toward the open ocean in the east

Formation Kear and Wood (1959). The post-shield erosion has resulted deeply incised relief, particularly northeast Upolu. These erosional processes formed amphitheatre-headed canyons, which merge to produce sharp peaks and a steep rugged terrain of narrow razor-backed ridges very typical of tropical erosion of volcanic islands across the Pacific (Figs. 2.12 and 2.13).

The Fagaloa Intrusion is strongly elongated from east to west and associated with columnar jointed basalts and a series of fractured coherent volcanic rocks of northeast Upolu. Like the older geological formation, younger rock formations such as the Salani, Mulifanua, Lefaga, Puapua, and Aopo are also associated with series of columnar joints in many parts of Upolu.

Fig. 2.12 Razor-back erosion of the lava shield of NE Upolu, as part of the Fagaloa Formation. (Photo taken from 13°55′4.69″S, 171°35′12.10″W)

Fig. 2.13 Razor-back morphology carved into the lava shield of Upolu. Fagaloa Bay is just behind the high peak in the middle of the view. View is looking roughly toward the NE

2.4 Geomorphology

Samoa Islands geomorpology is relatively simple. Here we use publicly available ALOS-PALSAR digital elevation models (DEM) and various terrain analysis techniques within a Geographical Information System (GIS) to extract morphological information from the available data. The terrain analysis is valuable approach to characterise the morphodiversity the islands hold. While this technique is powerful to gain new information on the landscape, it has also its own limitation largely associated with the resolution of the data. ALOS-PALSAR data sets are characterised with a 12.5 m spatial resolution that limit the geometry of the features can be recognized. This resolution however is good to recognized various characteristic linear structures, typical morphotypes of various, mostly volcanic, landforms.

The analysis of the morphodiversity started with the generation of various visual representation of the Samoa Island. This study resulted in the production of complex DEM with visual power the reader immediately can see the main morphostructural elements of the islands (Fig. 2.14). On the DEM the morphology of Samoa Islands is captured well and showing the younger morphological features in Savai'i (Fig. 2.14). The morphology of Upolu in the DEM also capturing the eroded nature of the exposed rocks, especially in the NE sector of Upolu. It is strikingly clear that the recognizable vent distribution resembles a very narrow, rift-like pattern in Upolu, that is almost exclusively east-west aligned within only 10 km band (Fig. 2.14). The vent alignments are clear in Savai'i as well, but there is another series of vents located mostly in a NE-SW trend that is not apparent in Upolu (Fig. 2.15).

Topographic cross sections also reveal the simpler topography of the higher and broader Savai'i against a more elongated and rugged surfaced Upolu as a reflection of the exposed erosional surface across Upolu (Fig. 2.15). The location of coral reefs is more common and continuous around Upolu, especially in the island's northern coastal areas (Fig. 2.15). Coral reefs are very restricted spatially around Savai'i and mostly restricted in the area facing toward Upolu and small sections in the northern regions (Fig. 2.15).

Fig. 2.14 Digital elevation model (DEM) based on ALOS-PALSAR digital elevation data showing a fundamentally shield morphology of both island that is more prominent in Savai'i

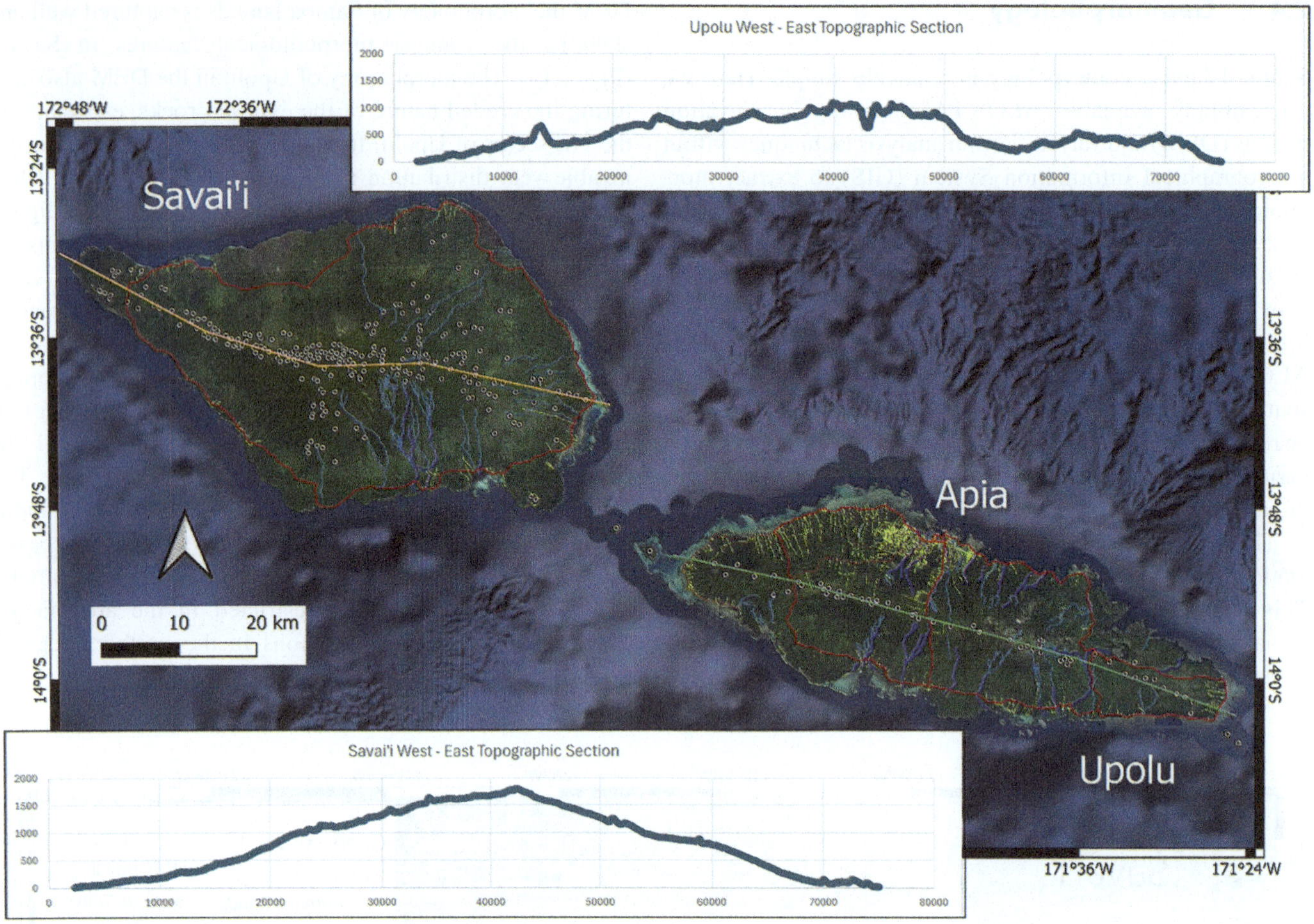

Fig. 2.15 Google Earth Pro satellite image of Samoa with the identified vent locations. Two cross section lines showing well the more rugged topography of Upolu indicating a post-shield erosion

Following the DEM-generation, the islands aspect ratio (Fig. 2.16), slope angle map (Fig. 2.17) and relief map (Fig. 2.18) was generated.

Aspect ratio map (Fig. 2.16) is particularly good to recognize structural settings by visualizing the main orientation large regions face toward. This research showed clearly the major ligaments located in the northern sector of Savai'i island and the older regions of Upolu where typical facets of erosional remnants of Pleistocene volcanoes form tropical erosion landscape and rugged topography, adding key elements to the understanding of the morphodiversity of the region helping to express the overall geodiversity. This study particularly valuable to pinpoint a major north facing collapse-like escarpment still not fully understood what origin it has. One of the most obvious interpretations is that it is a massive volcano collapse features that may occurred in the early shield building phase of the island, but due to heavy tropical rainforest coverage and young lava flow inundation, there is very limited surface exposures where mappable volcanic debris avalanche or other mass movement indicators. There is also no clear seafloor information in the northern sector of Savai'i, or at least the available resolution is inconclusive in this time. Interestingly extensive volcaniclastic breccias are mapped and known from the northern part of Upolu where steep and high hills with directional facets are visible on the aspect ratio maps indicating some sort of mass movement in the region but their nature weather, they are directly volcano collapse-related or more a proximal part of early volcanic edifice of the island is still under debate and investigation. From slope angle maps a simple slope classification map can be used to delineate those regions likely susceptible for landslides. For instance, slope angles over the angle of repose of loose materials are ready to move during intense rain and generate

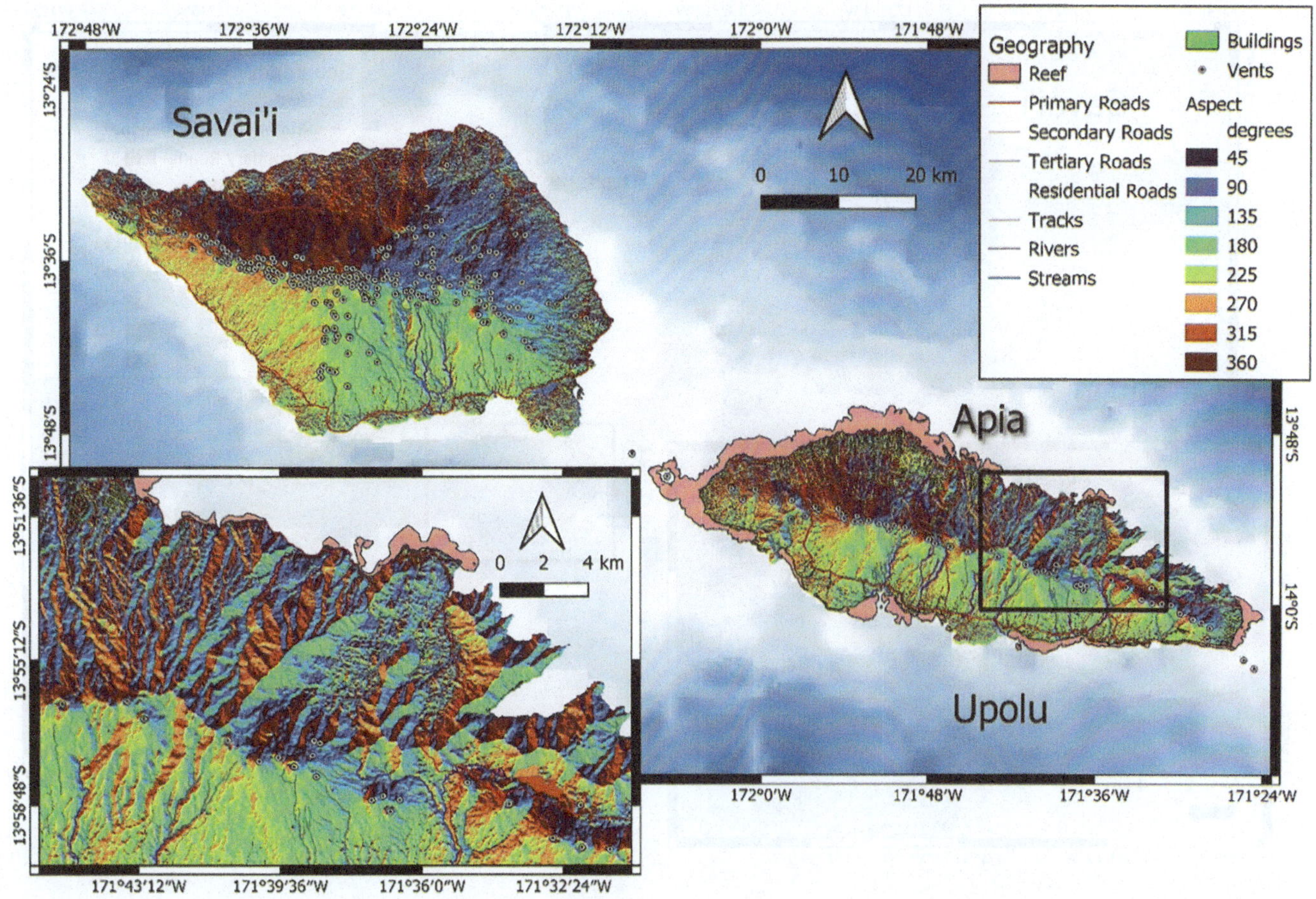

Fig. 2.16 Aspect ratio map of the Samoan Island. Note the very characteristic two-way facing slopes in Upolu, while in Savai'i more like four directions can be recognized. A close-up view of the Fagaloa Bay area showing very well the razor back hillsides as a reflection of tropical weathering of older lava shield of Upolu

landslide as well as large debris fans that can be channelized downflow into established fluvial networks (Fig. 2.18). The slope angle classification with the identification of areas where landslides can trigger mass movement and subsequently transporting debris toward the lowlands is a powerful tool to use it in geoeducation. The preservation of locations where both the landslide scars as well as their upper debris fan and distributed zones of debris aprons in low land are important geoheritage elements that can be utilized for geohazard education.

Drainage pattern of the Samoa Island also reflects well the geological architecture of the islands (Fig. 2.19). Based on the DEM, drainage pattern can be calculated that resulted a very good reproduction of the real drainage patter. The theoretical drainage network however is an indicative information on the potential areas where water can reach especially during intensive rainfall period, a scenario is common in the tropical condition.

The theoretical drainage network reflects well the underlying age of the substrate, having more complex networks in the old Fagaloa Formation territories (Fig. 2.19). The drainages are short and simple especially in Savai'i, while higher order drainage networks are more common on the erosional surface of the Fagaloa Formation, as a reflection of the post-shield erosional stage (Fig. 2.19). The catchment delineation also showed very clearly that Upolu particularly has narrow catchment areas running from the dorsal ridge to both side (north and south) of the island (Fig. 2.20).

Interestingly, the drainage basin pattern shows good correlation with the settlement boundaries indicating that the water accessibility must have been an important factor for the village communities to expand and evolve (Fig. 2.20).

In a valley depth map (Fig. 2.21) the distinction between Savai'i and Upolu is apparent. While Savai'i showing some trends of radial distribution of ridges and deeper valleys, in

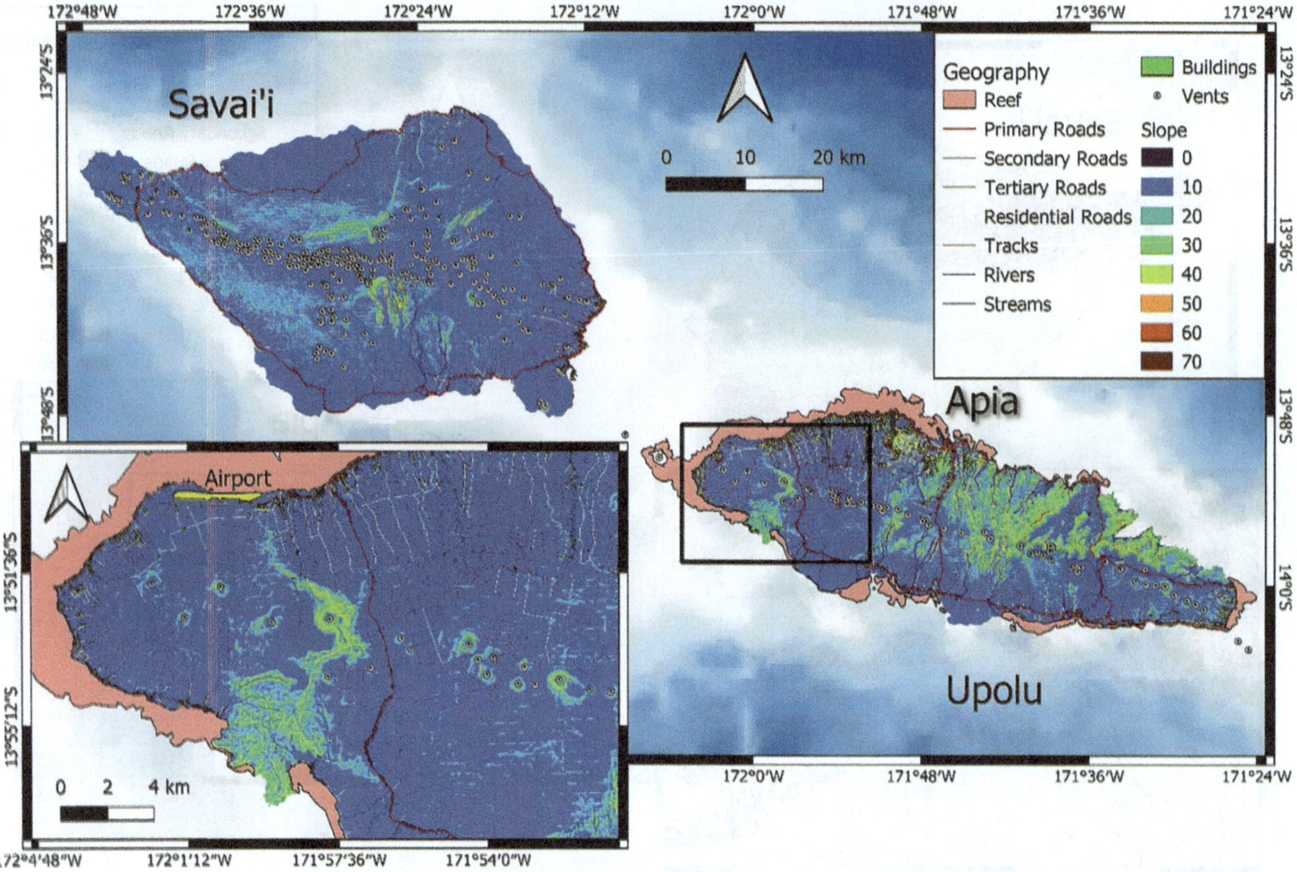

Fig. 2.17 Slope angle map of the Samoa Islands showing the rugged terrains of the erosionally carved northern Upolu. The slope angle map also showing very well the large island-scale volcanic scar in northern Savai' that is still under debate if it is truly representing a volcano collapse scar. The small inset map showing the western Upolu region where older lava flows characteristically visible as a rugged, steep sloped region where landslides in local scale can take place easily

Upolu the pattern is either two ways, to the north and to the south, while in the exposed post-shield surfaces it is very random (Fig. 2.21). Looking at the amphitheatre-like depression from Lemafa Scenic Site (Figs. 2.22 and 2.23) indeed suggestive to envision a caldera in the region, however, the geological setting is not supporting such model. Without further geological mapping in the region this question cannot be resolved.

In recent years and increasing number of new terrain characterisation method emerged. These methods are based on some mathematical algorithm to calculate to each location the variations of pixel elevations and visualise the results. While each of these generative methods has some advantages and draw back, one of the most widely applied landscape classification method is the so called geomorphon method (Fig. 2.24). A geomorphon is a representation of landscape based on elevation differences within the surrounding area of a target cell (https://pro.arcgis.com/en/pro-app/latest/tool-reference/spatial-analyst/geomorphon-landforms.htm#:~:text=A%20geomorphon%20is%20a%20representation%20of%20landscape%20based,within%20the%20surrounding%20area%20of%20a%20target%20cell). A total of 498 unique geomorphon patterns can be classified into 10 common landform types: flat, peak, ridge, shoulder, spur, slope, hollow, footslope, valley, and pit (Jasiewicz and Stepinski 2013). As geomorphons are derived from the digital elevation data, their texture will be heavily dependent on the resolution of the data

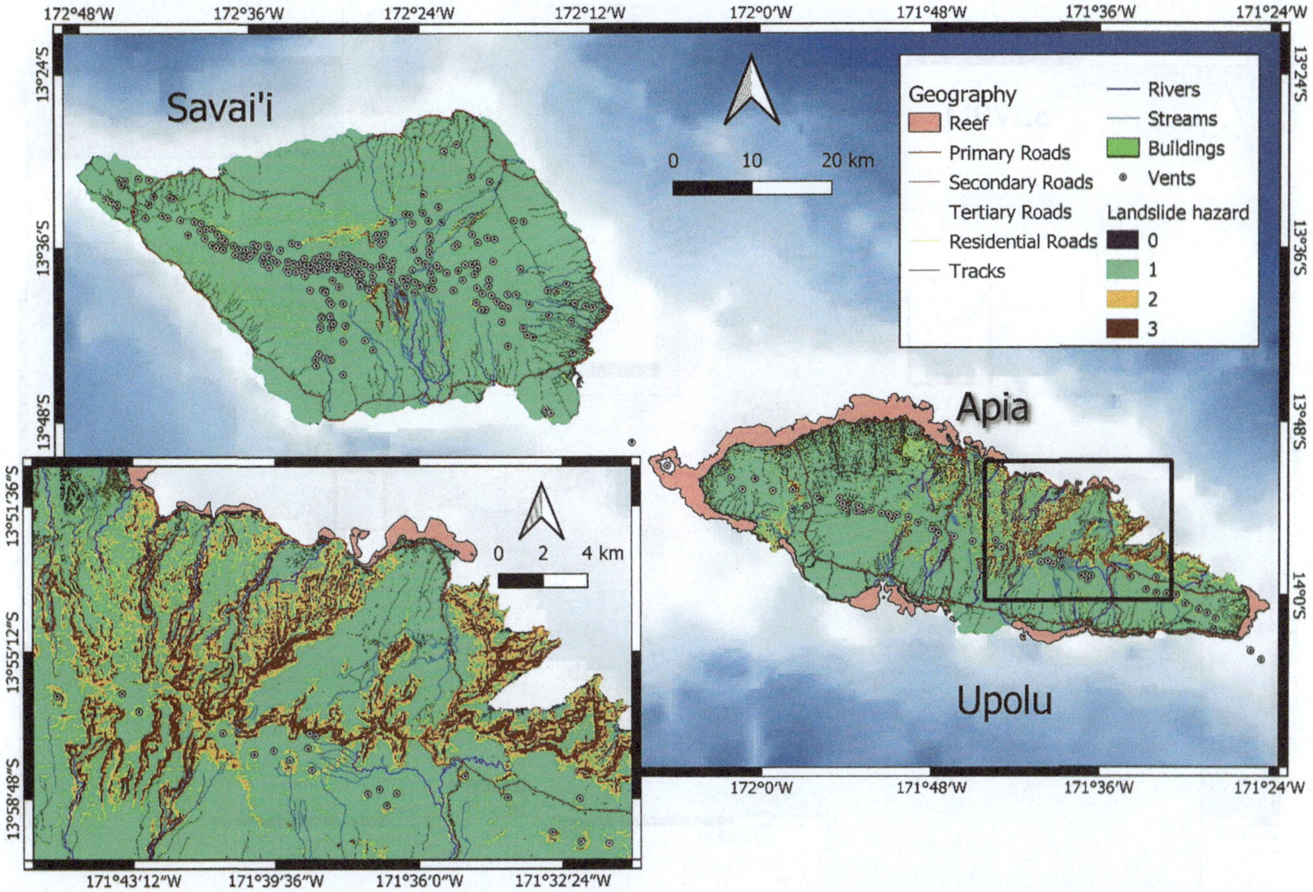

Fig. 2.18 Slope angle classification map as a proxy to identify landslide hazard zones and locations where landslide associated geosites maybe located. Such geosites are important elements to preserve and utilize in geoeducation purposes to develop a landslide resilient society. Slope angle classes are represented as 0 = <10, 1 = 10–20, 3 = 20–30, 4 = 30< in degrees. Note the steep and rugged terrain I the area Fagaloa Formation rocks crop out as part of the eroded lava shield volcano. Also note, that the slope angle category 3 and 4 areas occupy large areas that potentially can capture debris that can be transported toward the coastal settlement posing various level of flood hazard to the population. Such terrain analysis and the generated maps are excellent tools to visualize the hazardscape for the public through geoheritage

set. The DEMs used in this book are 12.5 m resolution suggesting that outcrop scale features can be captured by this method, but the data set is not suitable to design site specific geotrails or similar. Geomorphons however are easy to generate in any resolution from available DEMs, hence if LiDAR-based (**Li**ght **D**etection **A**nd **R**anging) data available, high resolution morphological classification of small terrains is also possible.

The geomorphon map of the Samoan islands are very informative (Fig. 2.24). The maps showing that the terrain dominated by slopes in both islands suggesting that the islands morphology is relatively simple, e.g., a large gentle sloping hill-dominated assemblage (Fig. 2.24). From this general trend areas where the oldest hence deeply incised valleys are located (in NE Upolu), the geomorphon analysis shows up deep valleys, ridges and occasional local depressions or peaks (Fig. 2.24). The geomorphon maps are very useful to identify location where outcrops hence likely geosites expected to be, and helpful to see the overall morphological architecture of the landscape. In various geodiversity estimates geomorphon concept has been utilized successfully.

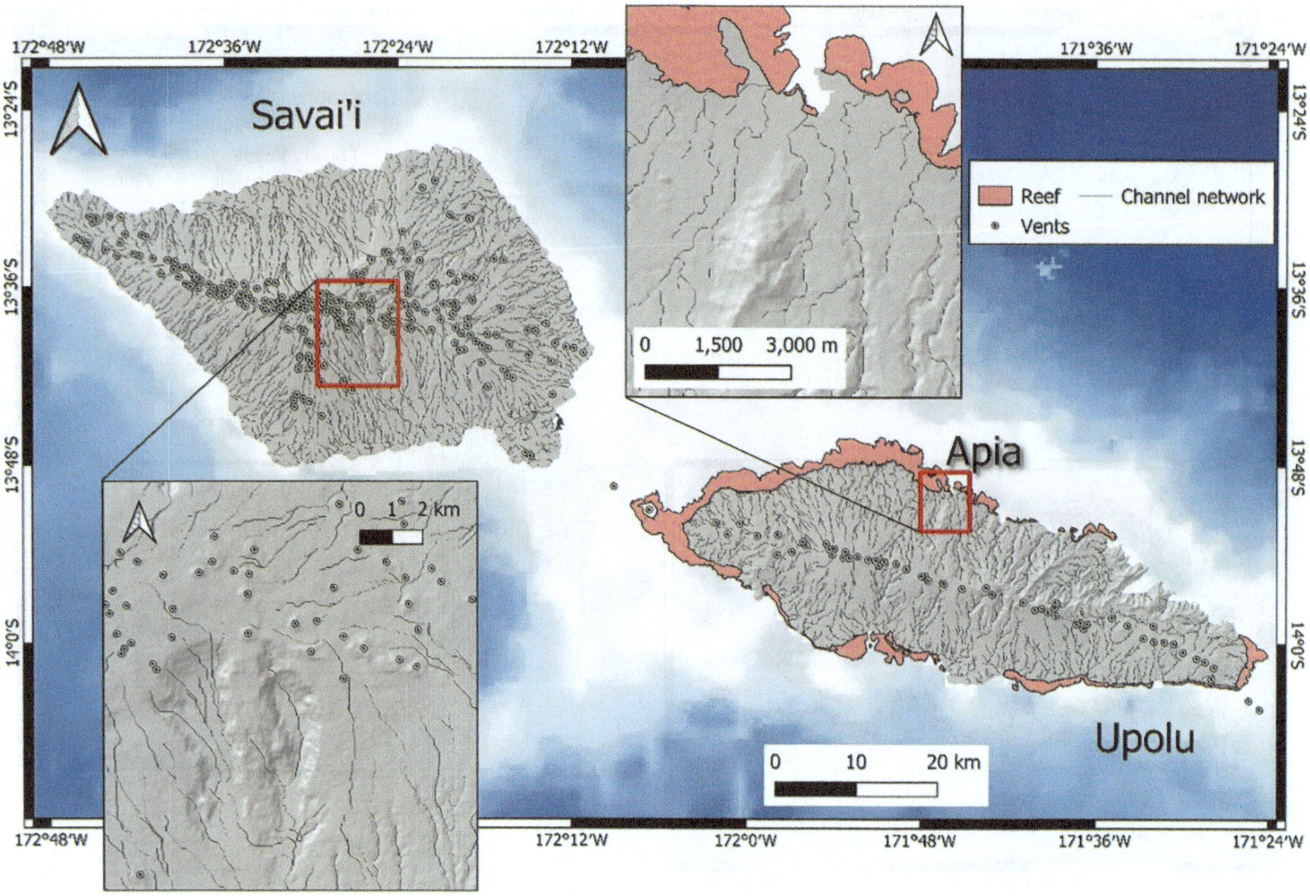

Fig. 2.19 Drainage network map calculated from the ALOS PALSAR DEM showing more mature characteristics for Upolu than Savai'i. The two inset map showing local features in drainage pattern sense from Upolu and Savai'i

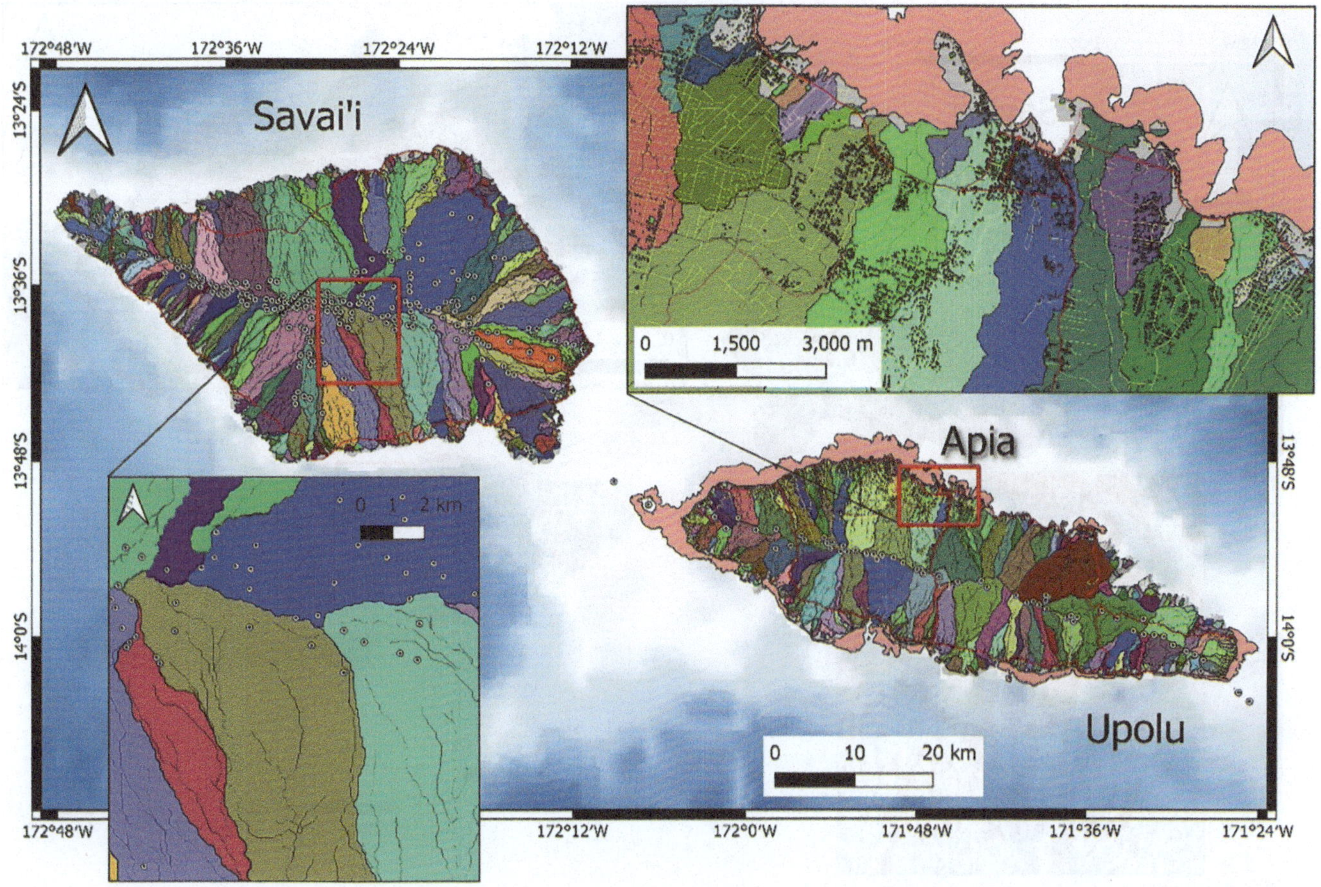

Fig. 2.20 Theoretical catchment areas of Samoa showing very narrow catchment holding short drainage valleys, typical for young terrains. The colours are only for visual representation without any specific relevance

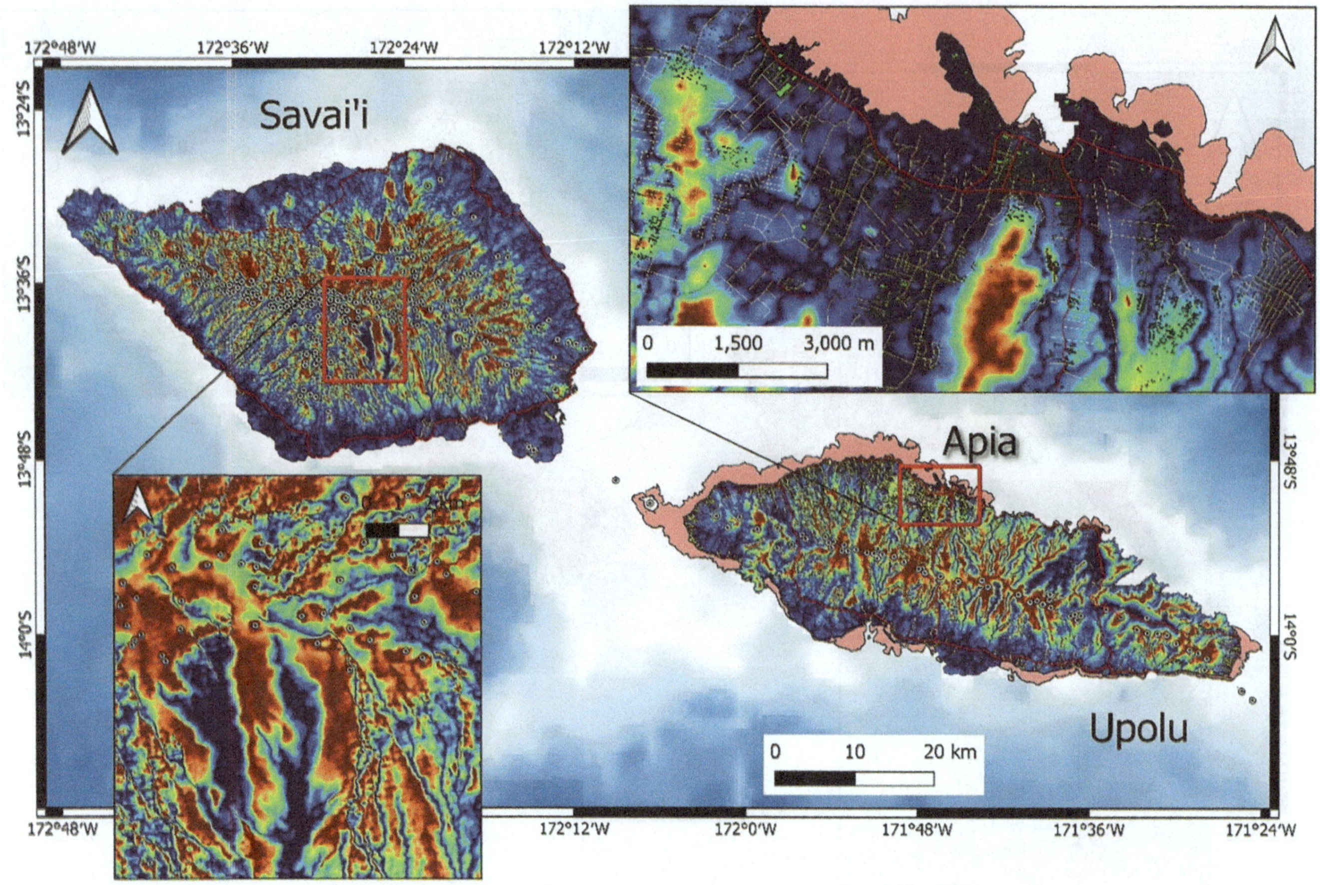

Fig. 2.21 Valley depth map (deeper blue is deep while brighter red is ridge-like) showing a characteristic distinction between Savai'i and Upolu. Also note the large blue areas as depressions around the Fagaloa region in the NE of Upolu, a feature commonly envisioned as a remain of old calderas. Norte the small inset showing the vicinity of Apia, a flat area connected to the dorsal rugged drainage network, a potential flood hazard zone in case of high intensity rainfall event

Fig. 2.22 Large amphitheatre-like morphology in the headland of the hillsides of the Fagaloa Bay area. Note the broad flat land that is open toward the ocean. View is looking toward NW from Lemafa Scenic Site along Richardson Road (13°57′20.88″S, 171°36′2.08″W)

Fig. 2.23 Landscape perspective on a Google Earth pro 3D satellite image looking from the Lemafa Scenic Site toward the coastal regions

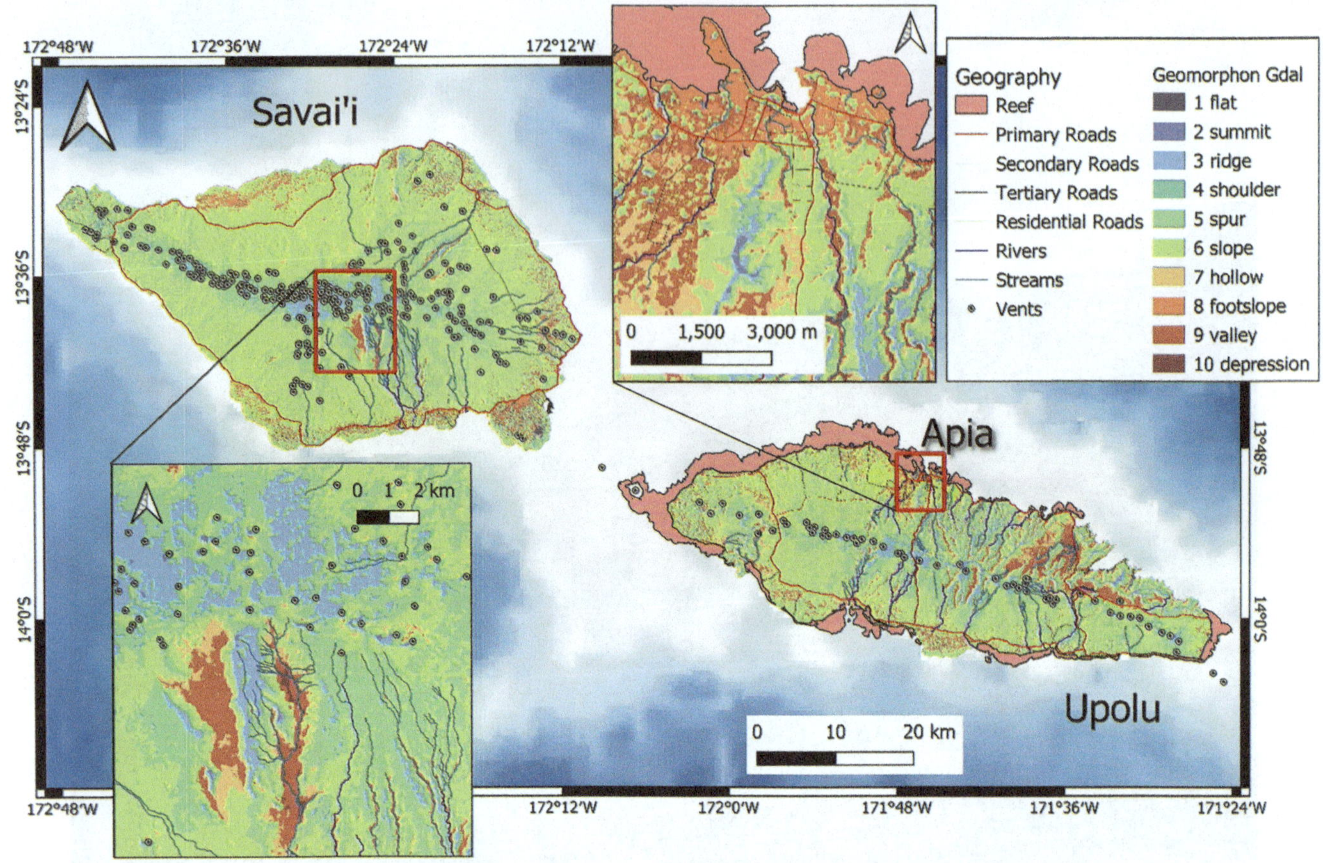

Fig. 2.24 Geomorphon map of the Samoan Islands. Note the dominance of slope areas in both islands. From this relative morphological monotony in the old regions, we can see valleys, and ridges that are associated with the location of outcropping Fagaloa Formation

2.5 Geohazards

2.5.1 Volcanic Hazard

The islands of Savai'i and Upolu are the western-most and largest islands of the Samoan group and have experienced a long period of volcanism, with the oldest deposits thought to be >2 million years (Kear and Wood 1959; Keating 1991). Volcanism on both islands appears to be focused along broad lineation's, which are commonly defined as rift zones or in some cases deep-seated faults probably associated with the crustal architecture beneath Samoa.

Although Upolu Island does not have the youngest of the mapped Holocene volcanic rocks present in the Samoan Group, it has several areas mapped as early Holocene in age (>10,000 years) and several volcanic features known that are fresh in appearance. These volcanoes are clearly sitting on the older successions, suggesting that they are associated with the rejuvenation of volcanism on the oceanic lava shield, potential 100s ka after the shield phase concluded. The prominent cone in western Upolu, named Tafua Upolu (Fig. 2.25), may be linguistic evidence for activity since people were present. The word Tafua is a cognate of Tofua, which means fire mountain—used for the active volcanic island in Tonga, and Tavuyaga (in archaic Fijian) used for a young volcanic cone on Taveuni, Fiji (Cronin et al. 2001). Notably the Tafua Savai'i in east Savai'i, is thought to be slightly older than 610 ± 60 years B.P. (Cronin et al. 2000). Other direct evidence of late Holocene volcanism on Upolu during the time of human occupation comes from a little-known published radiocarbon age of 1915 ± 65 years. B.P. (Grant-Taylor and Rafter 1962; Németh and Cronin 2009a, b). This date provides a maximum age limit on a tuff cone—or highly-explosive phreatomagmatic eruption from a vent just offshore of Cape Tapaga, east Upolu. Similarly—late Holocene volcanism has also been reported and described from Tutuila (American Samoa), farther eastward along the same major crustal structure (Keating 1991).

Eruption locations, as marked by lava-flow sources and volcanic cones are mostly concentrated along a broad east-west oriented zone, which is clearest on Upolu. On Savai'i the pattern of vent sites is more complex, with up to five rift-zones postulated by some authors (Kear and Wood 1959; Keating 1991).

Fig. 2.25 Prominent large, well-preserved scoria cone in the western region of Upolu, Tafua Upolu (13°52′40.51″S, 171°57′48.52″W) that is clearly sitting on an older lava flow field emitted from nearby vent locations indicating recurrent volcanism in the region, still likely within the rejuvenation period

Few recent detailed geological studies of Samoa have been undertaken. The most comprehensive geological mapping (at 1:400,000 scale) was conducted in the 1950s (Kear and Wood 1959). Little specialist volcanological work has been published or completed beside technical reports (Cronin et al. 2000, 2006) since and thus limited data exist to constrain age and frequency of past eruptions.

The type of volcanism on Upolu and Savai'i with recognizable volcanic landforms as part of the rejuvenated phase of the volcanic eruptive history of Samoa is known as "monogenetic" in the sense of their morphotypes (e.g., scoria and spatter cones and the erupted volume 1 km^3 estimated to each event), that is, each eruption location is normally only active once and the subsequent activity is located elsewhere (Németh and Kereszturi 2015). Hence, the islands are marked by hundreds of "one-off" volcanic cones or lava fields. Lava flows however in the bulk volume of the islands in stratigraphically basal position part of the lava shield, while the capping flows are associated with localized cones along the major rift axis of both islands. The implication of the monogenetic nature of the most recent volcanism is that it is almost impossible to predict the location of the next eruption site with certainty, especially without reliable and robust age data set that is treated statistically and confirmed by volcano geology studies.

Due to the uncertainty and the low number of direct age dating accompanied by the suspected young (Late Pleistocene to Holocene) eruption ages as well as the "one-shot nature" (monogenetic) of the identified vents justify a volcanic hazard estimation based on various eruption scenarios the geological architecture indicates. Such scenario-based approach also easy to be adapted into geoheritage research and geosite localization works to provide a robust evidence-based data set to delineate the upper and lower limits of volcanic eruptions, their impacts and potential mitigation strategies. Geoheritage in this conceptual framework can play vital role as it can deliver stories to the communities that can be visualized and tested in respect of potential responses of specific communities. This approach also can provide great avenue for community-based geoeducation strategies where co-development and co-design can be linked strongly to the inventoried geoheritage elements.

The zonation of potential vent locations is derived from the identification of vent sites (e.g., centre-points of recognizable volcanic edifices) on satellite imagery, aero photographs and various geomorphologic inferences of the underlying structural geology of the islands derived from direct field surveys and desktop works of available data. The most obvious features are the distribution of volcanic cones. The vent distribution map based on simple heatmap without consideration the actual age pattern of the identified vents if all the vents recognizable represent young, post-Pleistocene eruptions, hence the events can be considered as part of the current state of the magmatic system. The vent density maps for the entire Samoan Islands (Fig. 2.26) with focused maps for Upolu (Fig. 2.27) and Savai'i (Fig. 2.28). The vent density map is clearly indicating the more irregular pattern of young volcanic events in Savai (Fig. 2.26).

For the lava inundation lava flow simulation models such as the Q-LavHA can be used (Becerril et al. 2021; García et al. 2024; Hakim et al. 2022; Mossoux et al. 2016; Németh and Moufti 2023; Prieto-Torrell et al. 2021; Vilches et al. 2022), however, their result needs to be treated with caution in case of flow fields with low viscosity magmas that tend to be dominated by pahoehoe lavas and form lava tube networks in Samoa. In both islands as we don't have strong case to delineate the most likely vent locations, the current high vent density areas as well as the information about the location of the youngest vents been used.

In Upolu the actual vent location is difficult to identify as the vent density map shows few higher density roles without any systematic age progression. In this regard the modelling was focused on the hazard management questions, what if, a new vent would open in the hill above Apia, in the region near the airport in the eastern dorsal ridge where some data shows young eruption ages (Fig. 2.27). The modelling used Manhattan method that is preferring lava propagation along channels (e.g., within preexisting valleys) as Upolu has a more pronounced erosional network upon the young volcanism operated. The lava flow run out distances were fixed to be 15 km that is the longest known lava flow length. The flow thickness was fixed to be 10 m with a buffer of 20 m allowing the model to overcome local errors on the coarse resolution DEMs. The models showed very clearly that lava flows, even in large volume would not reach the downtown region of Apia but could make significant inundation in the first morphological terrace defined as the erosional surface of the shield volcano (Fig. 2.27). In western Upolu however, lava flow inundation that could even reach the current airport is possible (Fig. 2.27).

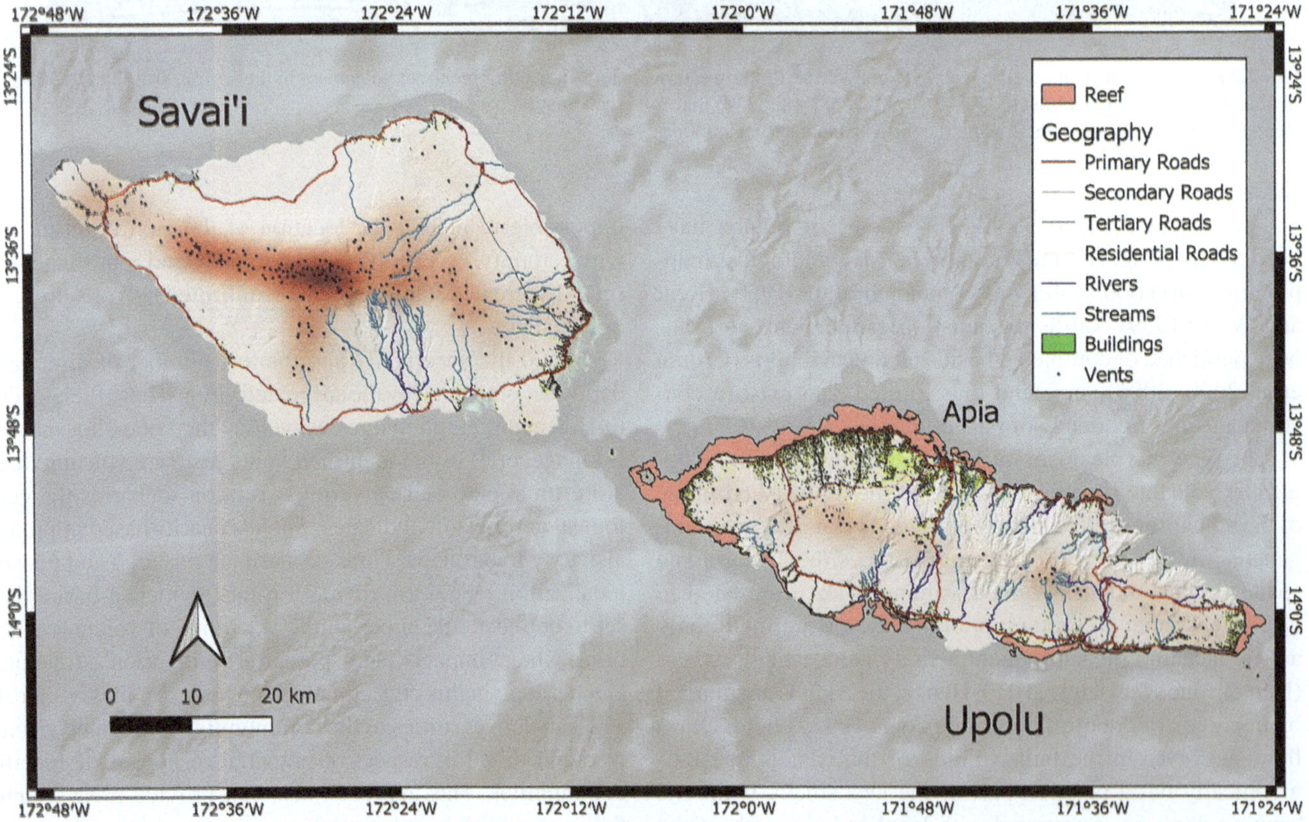

Fig. 2.26 Vent density map showing distinct pattern in Savai marking multiple rifts, while in Upolu the vents are more regularly aligned following the NW-SE dorsal range

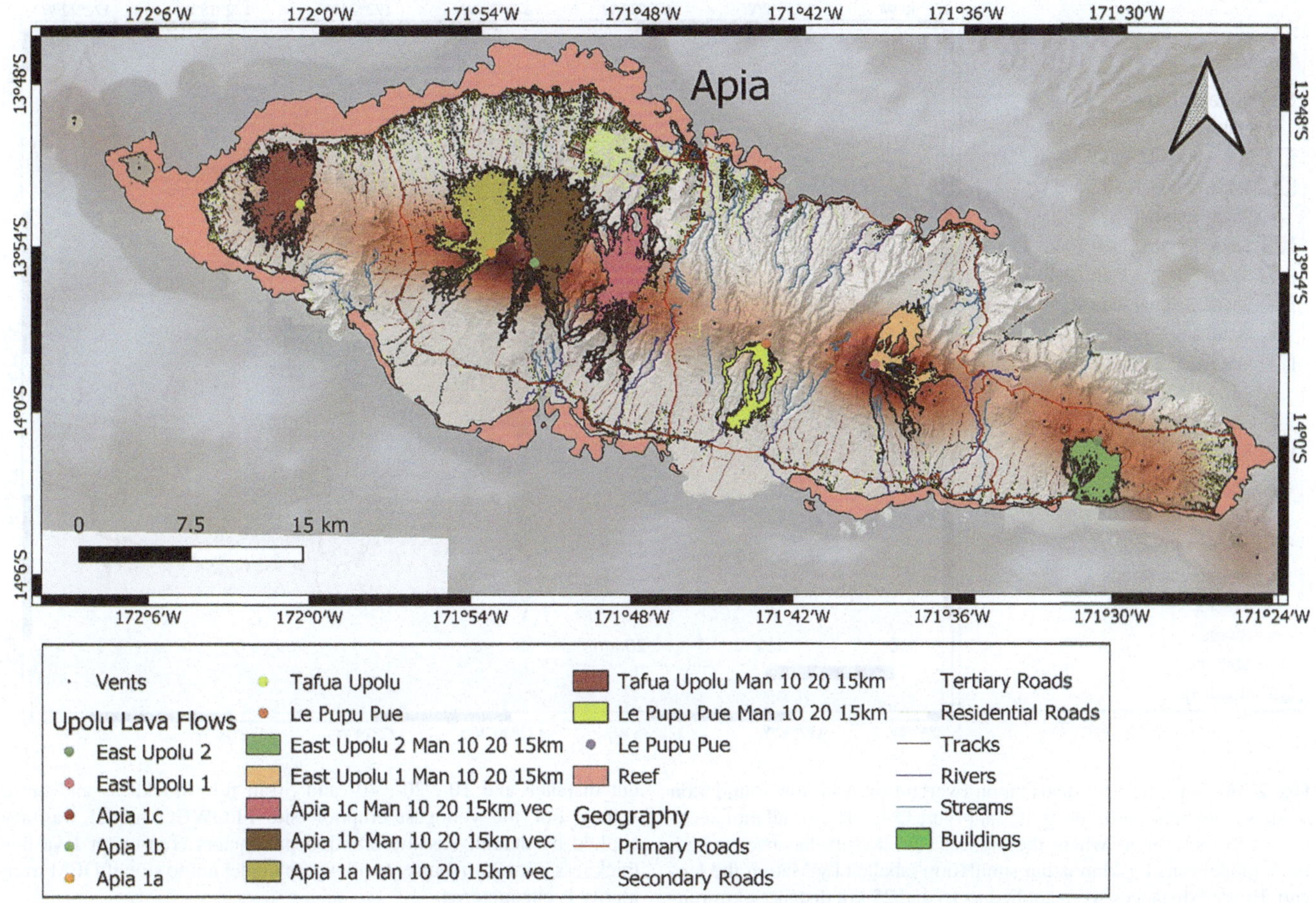

Fig. 2.27 Vent density map of Upolu showing darker red zones along the dorsal rift where volcanoes erupted. The map also showing lava flow inundation models applying Q-LavHA simulation techniques to those areas where the highest vent density has been noted. Each model runs by Manhattan simulation, 15 km runout distance and 10-m lava flow thickness with 20-m buffer to allow the model not to stop if DEM irregularity is encountered

In Savai'i potential future eruption sites are assumed along the major rift zones, especially in those areas where Holocene eruptions are known (Fig. 2.28). Vents are around the central areas of the dorsal ridge facing toward the north side of the island were selected as potential future eruption sites for the simulation.

In Savai'i the simulations used a bit larger value based on the preliminary mapping of lava flow fields and their onsite characteristics (Fig. 2.28). From the summit region where historic eruptions initiated a runout distance of 15–25 km was selected with Manhattan simulation with lava thickness ranging from 10 to 40 m with an additional 10 m for buffer purposes. Interestingly the lava flow simulation had difficulty to reproduce the lava flow distribution pattern currently seen in the field having lavas all the way down to the current coastline. The reason for this is likely that the lava flows have significant tube network where 10 m across lava tubes potentially acted as established near-surface melt arteries that were delivering lava far away, preserving its mobility. Savai has some spectacular lava tubes that are traceable of 100s meters and showing clear evidence on their very dynamic nature (Figs. 2.29 and 2.30).

These areas of cone development mark zones of weakness and faults where magma has risen in the past (Walker 1986, 1987). The cones are not an absolute guide, since many lava-field eruptions may just form from fissures, leaving little geomorphic evidence of the vent location. In addition, some of the cones/vents may form downslope of the original rise point of magma—with magma transferred laterally through subsurface rift zones—such as on Etna volcano or Hawai'i (e.g., Kilauea Volcano). Lava tubes also play significant role in melt transportation in hidden subsurface network as it has been documented in other large lava-dominated volcanoes like Jeju Island in Korea (Jeon et al. 2024; Kim et al. 2022), or Etna in Sicily (Calvari et al. 2024; Calvari and Pinkerton 1999; Spampinato et al. 2024).

Identifying more specifically the potential location of volcanic eruptions inside these zones will only be possible by application of a seismic network that is able to record local seismicity maybe associated with magma movement.

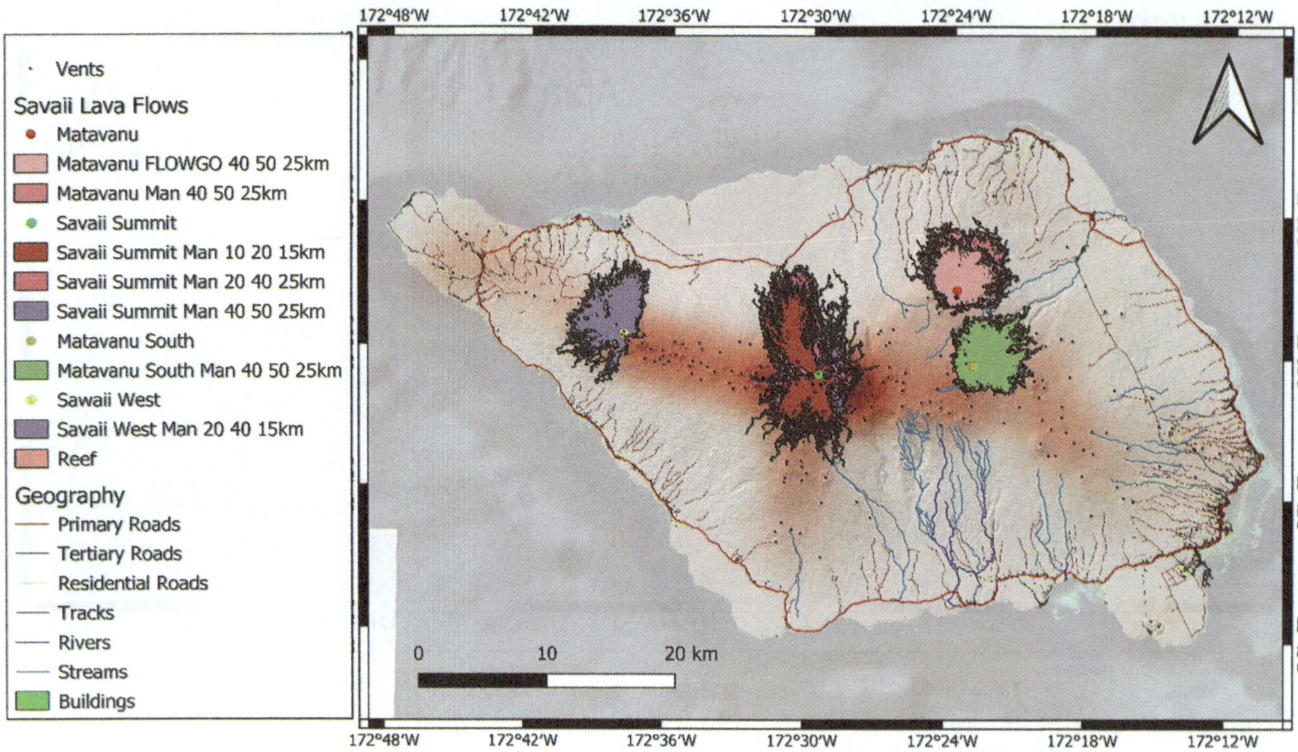

Fig. 2.28 Volcanic vent distribution overlaid by lava flow inundation models from preset eruptive points applying Q-LavHA simulation techniques to those areas where the highest vent density has been noted. Each model runs by Manhattan simulation labelled by Man in the legend. Runout distances were applied to 15 and 25 km distances km runout distance and 10-, 20-, 40- and 50-m flow thickness and buffer values. For the youngest eruption site FLOWGO model was also applied by default basaltic eruption parameters. Ten-meter lava flow thickness with 20-m buffer to allow the model not to stop if DEM irregularity is encountered

Currently a robust volcanic hazard scenario documentation and communication of its meaning to the community is probably the best possible strategy for volcanic hazard management. In this context volcanic geoheritage is an important avenue for volcanic hazard resilience development.

A typical volcanic eruption events can be distinguished of several phases. In these phases different eruption styles, processes and impact can be recognized. Each of the recognized volcanic processes such as lava flows, ashfalls, pyroclastic surges, littoral explosions as the most likely expected eruption scenarios for Samoa poses different kinds of hazards to the community. However, rather than considering each volcanic process in isolation, it is more useful to examine what range of processes would occur in typical volcanic events. Each of the recognized volcanic eruption or volcanism-related event scenarios has their unique geoheritage elements. Therefore, the eruption scenarios identified based on the preserved volcanic successions of Samoa can directly be linked to their geoheritage elements. In Samoa, five type of volcanism related event scenarios have been distinguished (Table 2.2) such as (1) long-term, lava field eruptions, (2) short-term spatter-cone eruptions, (3) explosive phreatomagmatic or hydrovolcanic eruptions, (4) mild explosive scoria cone eruptions and (5) submarine flank collapse. The lava flow effusion is probably the most common and expected hazard type as lava flows are associated with most of the known Pleistocene-Holocene volcanoes. They are typically low viscosity, free-flowing pahoehoe lava types with lava tube networks (Cas et al. 2024). In some cases, lava flows are associated with short fissure aligned lava spatter cones or fed from larger volume scoria cones typically formed by magmatic volatile-driven mild explosive eruptions (Giordano et al. 2024). In case of external water availability, magma-water explosive interaction can form tuff rings or tuff cones depending on where the vent opens (Németh and Kósik 2020). While not fully demonstrated, the volcanic flank collapse cannot be neglected as a potential volcanism-related phenomena in Samoa. Each of these volcanic scenarios has distinctive geoheritage elements (Table 2.2).

Fig. 2.29 Pe'ape'a Cave along the northern main road of Savai'i. The lava caves are important habitats for swallows. The cave can be walked over 100s of meters (13°30′23.17″S, 172°28′40.56″W)

2.5.1.1 Long Lava Flow Effusion

The long lava flow scenario best preserved example is the 1905–1911 Matavanu eruption (Table 2.3). Many descriptions of parts of this eruption and its effects are available (Anderson 1910). As a recognition of the eruption site geological heritage significance the Matavanu eruption site has also been selected to be listed in the International Union of Geological Sciences First 100 Geological Site in 2022 (https://iugs-geoheritage.org/geoheritage_sites/1905-1911-matavanu-volcanic-eruption/). These descriptions give rise to a scenario of events and hazards (Table 2.3).

The eruption site has a n extensive near-shore young lava field (Fig. 2.31) with spectacular pahoehoe surface textures (Fig. 2.32). In addition, the eruption's long pahoehoe lava flow destroyed the London Missionary Society Church of Saleaula that stands as a memento of the scale of impact such eruption can generate (Figs. 2.33 and 2.34).

2.5.1.2 Short-Term Spatter-Cone Eruptions

This style of eruption may precede the major lava flow events such as the 1905–1911 Matavanu eruption. Source regions of long lava flows can be limited lava spatter cones or ramparts, but in many cases, these are eruptions can only be of volumes of magma-fed events. Many cases long lava flows may even have no significant and easy to identify vent sources (in many cases due to heavy vegetation they may be hidden). Many examples of these kinds of events are present on Savai'i and Upolu, that can be easily accessed such as those in the Seuseu area of West Savai'i (Figs. 2.35 and 2.36). The 1902 eruption on Savai'i, sourced from Mount Mata ole Afi (13°36′29.19″S, 172°30′27.77″W) that initiated from the dorsal range of Savai'i is a good example of this type of volcanic hazard, however the location itself is very inaccessible (Fig. 2.37). From this data a scenario with the potential geoheritage elements can be summarised in Table 2.4.

2.5.1.3 Hydrovolcanic Explosive Eruptions

These styles of eruptions are arguably of greatest potential hazard to life in Samoa (Cronin et al. 2006). Due to their rapid acceleration to explosive phases (Németh and Kósik 2020), they have the potential to catch local populations unawares. As magma—water explosive interaction is more expected in water-rich lowlands such as coastal plains (Agustín-Flores et al. 2014; Ilanko et al. 2024; Kereszturi et al. 2014; Németh et al. 2012; Németh and Kósik 2020) where much of the population resides currently, this type of volcanic hazard poses significant risk to the human society hence every strategy to be the community aware of it is important. Population prior the pre-European contact time perhaps was much larger inland (Jackmond et al. 2018) due to avoidance of malaria and coastal flooding, humidity still fishing and seafood gathering always connected Samoans to the coastal regions where explosive hydrovolcanic is a likely volcanic eruption scenario. Coastal settlements also played crucial role for human occupation development of Samoa (Kane et al. 2017). Geoconservation of such volcanic geology sites therefore important to preserve them for geoeducation purposes. Also, these types of volcanic events are far more explosive than the eruption driven by the magmatic volatiles of the mafic basaltic magma (Németh and Kósik 2020). It is well known that the interaction of magma and water induces a highly explosive fuel-coolant interaction (Németh and Kósik 2020). This generates highly energetic eruptions with significantly larger risk footprint than a same volume magma can generate without interacting with external water (Németh et al. 2012).

Fig. 2.30 The interior of the Laauoleola Cave near Aopo A'opo (13°31′49.72″S, 172°31′14.22″W) in the northern road of Savai'i as part of a lava flow field formed about 300 years ago. Note the lava marker horizons and the pavement of the free-flowing flow system. Person is 168 cm tall

Typically, there is an ideal ratio of magma to water to achieve the maximum energy release (hence destructive power) of such interactions (White and Valentine 2016; Wohletz 1983, 2002; Zimanowski et al. 2015). Too much water dampens explosivity, while too little results in also a lower-energy eruption and transition to more normal magmatic explosive or lava-flow eruption. The other feature about these eruptions is the generation of so-called pyroclastic surges (Houghton et al. 2015; Roche et al. 2009) that are rapidly expanding mixtures of gas/ash/steam that blast outward from the eruption site along the ground or water surface. These are rapidly moving (≫30 m/s) and destroy most things in their path. They offer the greatest danger to life, of all the recognized Samoan eruption styles. A good example of this type of event is demonstrated in the well preserved, accessible pyroclastic successions of the Tafua Savai'i volcanic complex in eastern Savai'i (Fig. 2.39). Apolima Island itself is a fine example of a tuff cone (White et al. 2015) that is under wave cut erosion been cut back to its core and within its crater a small community lives now (Fig. 2.40). This volcanoes are typical shallow marine features where the initial explosions took place below the shallow water table, but pierced the water surface and gradually emerged from the sea such as it was observed during the 14 November 1963 Surtsey eruption offshore of Iceland that gave its name for this type of volcanism as Surtseyan (Cole et al. 2001; Thordarson and Larsen 2007). Similar rift-edge tuff cone remnants of the small offshore islands in the eastern edge of Upolu (Németh et al. 2007). While there is little evidence to support magma-water explosive eruption styles inland of the main islands, in the regions closer to the sea level (less than 200 m) in the eastern region of Upolu there are some morphological evidence of existence of fine ash that eroded characteristics gully network on large but broad cones, indicative for potential magma—water explosive interactions (Fig. 2.41). The typical hydrovolcanic explosive eruption scenario (Table 2.5) is based upon a geological interpretation of the Tafua-Savai'i complex (Cronin et al. 2001); the recent Ambae eruption (Németh et al. 2006); and other historical examples of these types of events worldwide (Figs. 2.42, 2.43, 2.44, 2.45, 2.46, 2.47 and 2.48).

Table 2.2 Potential eruption types on Upolu and Savai'i. Each of the mentioned eruption scenarios has numerous locations where their eruptive products are still preserved and should be incorporated in geoconservation programs to be able to use it for geohazard education

Eruption/event type	Examples	Duration	Area fully destroyed	Area partly damaged	Potential human impacts	Geoheritage/geoconservation
1. Long-term, lava field eruptions	1760, 1905–1911	Many years, can be decades	10 to >200 km^2	<10 km^2	Low potential for loss of life, but permanent damage to property and farmland	Lava flow surface textures in Savai'i's 1905–1911 Matavanu eruption site OR lava flow surface textures at the Holocene Le Pupu Pue lava field. Local lava flows can form local geosites that local communities can utilize form their geoeducation purposes as knowledge preservation
2. Short-term spatter-cone eruptions	1902, Seuseu W-Savai'i area	Days to weeks	2–10 km^2	10–50 km^2	Low potential for loss of life; localised damage to property and farmland	Specific lava spatter fields in western Upolu, just east of Tafua Upolu preserved vegetated lava spatter cone fields that are still accessible. In high ground fissure aligned lava spatter crater rows of the 1902 Seuseu region in Savai'i. Each island has large number of such small spatter cones with short lava flows that can play significant role of knowledge preservation and awareness raising hence localization and inventory of such sites could form community geoconservation goals
3. Explosive phreatomagmatic or hydrovolcanic eruptions	Tafua Savai'i	Weeks to months	10–30 km^2	50–100 km^2	Moderate potential for loss of life, localised damage to property and farmland	Tafua Savai'i coastal tuff ring section or the eastern Upolu offshore islands are those where well exposed hydrovolcanic pyroclastic successions preserved. Apolima Island itself is a unique geosite with accessible outcrops to see the results of explosive magma and water interaction
4. Explosive scoria cone eruptions	Tafua Upolu	Weeks to years	10–30 km^2	50–200 km^2	Low potential for loss of life, localised damage to property and farmland	Large scoria cones such as Tafua Upolu in Upolu, or the scoria cone rows in eastern Savai'i are good candidates to show the geometry and impact of such explosive eruptions
5. Submarine flank collapse	S-Savai'i	Several hours	Unknown	Unknown	High potential for deaths through associated tsunami	Mostly the natural environment of the steep escarpment of northern Savai'i can be conserved as a location where people can see the morphological impact and its scale of a potential volcanic landslide

Table 2.3 Summary eruption scenario of the 1905–1911 eruptions derived from volcanic scientific literature and historic reports

Timeline	Events	Impacts/hazards	Actions required	Geoheritage/geoconservation
0–10 days	Earthquakes detected on seismographs—very few felt by people	None	Identification of eruption zone. Public warning. Installation of volcanic monitoring network—seismic	Some cracks on the London Missionary Society Church of Saleaula are likely the result of such precursory events
10–20 days	Eruption starts, producing small-scale explosions to build up one or more spatter cones	Vent area—burial of c. 200 m radius by spatter cone. Volcanic bombs thrown out to 1.5 km radius—common in 1 km from vent Ashfall several cm thick in areas up to 3 km away	Evacuation of people in a 3 km radius of vent. Evaluation of lava flow pathways and potential new hazard areas. Evaluation of wind direction and ash-fall hazards. Public information Regular aerial monitoring of eruption	The Matavanu scoria cone with its deep crater is a geosite where the source vent of a long-lived lava flow effusion site can be observed. This location requires high geoconservation status

(continued)

Table 2.3 (continued)

Timeline	Events	Impacts/hazards	Actions required	Geoheritage/geoconservation
20–30 days	Lava flows begin—either from an opening of the crater (wind direction or collapse influenced) or from the base of the cone	Burial by lava of areas up to 3 km from source Minor or no ashfall	Evacuation of lava flow pathway and a down-slope projection from the pathway up to 10 km from the flow front. Regular aerial monitoring to predict lava paths	Lava flow can be traced from source to the unconfined lava fan. The transitional zone has high geoheritage values as likely contain lava tubes that helped effectively transport lava from source to the unconfined lava field. The Matavanu Crater is an accessible geosite to see the result of a long-lived lava flow emitting volcano that created a deep pit crater within lava spatter layers exposed
30–90 days	Lava flows reach the coast on several fronts, begin filling lagoon areas	Burial by lava of major areas, >10 km from source. Littoral explosions where lava traps beneath in lagoons—exploding blocks to up to 100 m from explosion site. Production of sulphur dioxide (SO_2) gases and aerosols—known as vog (volcanic fog)	Evacuation of all lava pathways and areas within 1 km of the lava entry to the sea. Monitoring of wind direction for volcanic gas/vog emissions. Regular aerial monitoring of flow entry points and flow margins for breakouts	Costal littoral cones are preserved within the lava field, and they have high geoheritage values and geoconservation status. They are accessible by easy walk on the lava surface and can be used for geoeducation to demonstrate what could happen if lava enter the ocean
3 months to several years	Lava flows continue underground through tubes and exit at the sea Large lava blisters may form near coast and collapse suddenly Lagoons fill and lava spills over reefs Lava platforms collapse into the sea Eruption intensity waxes and wanes—with periods of several months having little activity, followed by resurgence of activity	Burial of new areas by lava as flows break-out from sides of main channels. Tsunamis generated in areas up to 10 km of lava ocean entry as lava platforms collapse into the sea. Vog of variable intensity dispersed with wind—crop damage. Minor ashfalls up to 5 km from vent and from ocean entry of lava on rare occasions—crop damage and respiratory impacts	Evacuation of all areas up to 1 km from sides of lava flows. Regular aerial monitoring of flow propagation and potential lateral breakouts. Continued volcanic monitoring with seismographs and wind/weather monitoring for vog/ash forecasts. Evacuation of near-coastal areas within 10 km of lava entry point to protect from tsunami hazard Regular public briefings	The main part of the pahoehoe lava flow field with all the key features associated with pahoehoe lava flows accessible easily. Geosites can include inflation features such as updoming and roof collapse, fractures, and lava plates displaced and local tumuli
After 1 to several years	Eruption slowly diminishes in intensity, lava flows diminish, lava tubes empty, vog emission decreases and eventually eruption gradually ceases. After 12 months of no activity, eruption has finally ceased	Reduced intensity impacts of hazards from all types, but possible resurgence of activity for short bursts	Ongoing actions as above, winding down following 12 months of inactivity	Deflation features and lava drainage locations. The post-eruptive pioneer vegetation demonstrates well how a lava inundated area could recover after an eruption

2.5.1.4 Scoria Cones

The explosivity of all eruptions depends in the first instance on the gas content of the magma when it reaches the ground surface. For the basaltic magmas typical of Savai'i and Upolu (Kear and Wood 1959), gas escape occurs well before the magma reaches the surface, hence the eruptions are typically of low explosivity—spatter eruptions and lava flows. In some cases, however, with more evolved magmas, where viscosity is slightly higher (e.g., trachy-basalt and trachyte; Kear and Wood 1959), gas contents are retained more efficiently and more explosive eruptions result. These most explosive "dry" style of eruptions in Samoa generates large cones of scoria and more widespread ashfall hazards—a good example being the apparently young Tafua-Upolu. These events can

Fig. 2.31 Pahoehoe lava flow field of the 1905–1911 Matavanu eruption (13°27′34.29″S, 172°19′9.54″W)

be also relatively long lived—several years or more. Similar types of eruption are typical on Taveuni island in Fiji (Cronin and Neall 2001). The following scenario (Table 2.6) is based on the famous historic eruption of the Paricutin volcano in Mexico, a volcano witnessed from birth onward as it grew from a farmer cornfield (Luhr and Simkin 1993). In addition, subsidiary information comes from the geology of Samoa and other similar eruptions worldwide (e.g., Self et al. 1974).

2.5.1.5 Submarine Flank Collapse

A common feature of large oceanic islands such as Savai'i and Upolu are that they rise steeply from surrounding deep ocean floors and they also exposed to strong wave cut erosion. In addition, their slopes are made up of lavas and primary and reworked volcaniclastic material of very fragile, commonly altered and rheologically heterogeneous rock assemblages that is also sitting over unstable, mud-rich marine sediments. Due to this geological setting, oceanic volcanic islands are expected to be unstable landforms that can deform, or even collapse into the marine basin (Biain et al. 2016; Coello-Bravo et al. 2020; Groom et al. 2023; Idárraga-García et al. 2021; Santos et al. 2019; Schmincke and Sumita 2016). The gravitational instability of volcanic islands can also lead to a gradual spreading of the islands without major mass wasting phases just leaving behind complex geological structures within small areas. Large scale volcano collapse in ocean island environment can also cause mega-tsunami hence can cause significant hazard for the coastal societies (Paris et al. 2017; Ramalho et al. 2015). In several other similar volcanic chains, such as Hawai'i and the Canary Islands, such large-scale collapse events are known or are postulated to have formed major inter-regional tsunamis that are suspected to be the result of major volcano collapse occurred in the history of the volcanic islands (Carracedo 1999, 2014; McGuire 2006; Pérez-Torrado et al. 2006). It is also expected that major collapse events likely fed massive submarine landslides hence volcanic debris avalanches, rock falls, or regular volcanic debris flows together are the likely geological phenomena associated with volcanic islands mass wasting (Boulesteix et al. 2013; Masson et al. 2008). Large scale volcano collapse features and their resulting debris-avalanche deposits have also been recognised on the submarine slopes of Savai'i and Upolu and suggested to be associated with major volcano collapse events within

Fig. 2.32 Pahoehoe surface textures of the 1905–1911 Matavanu eruption lava flow field (13°27′31.59″S, 172°19′1.66″W)

the shield building phase of the volcano island evolution (Hill and Tiffin 1993; Keating et al. 2000). The prediction of such large events is currently impossible. The main hazard from submarine landslides is that of tsunami. Some authors have estimated run-ups of large Hawai'ian events as up to 300 m in the nearby areas (Moore et al. 1989). Certainly, many coastal villages that are often <10 m above sea level would be destroyed. We currently have no idea of the frequency of such events in the Samoa area, and none have occurred from any of the world's oceanic islands in recent history. From geoheritage perspective, only the northern escarpment of Savaii as a major structural element can be noted and inventoried as a geosite however it is not so characteristic feature from the ground. Its scale and geometry can only be appreciated from various air photo or satellite imagery services.

2.5.2 Landslides

Landslides are caused by rain, earthquakes, volcanoes, or other factors that make the slope unstable. Hence landforms that are weaken due to alteration or weathering are more susceptible for landslides (Fig. 2.49). The original geological architecture, such as steeply inclined beddings, faults or tight folds with contrasting rheology rock assemblages are also can be the primary reason for landslide susceptibility (Fig. 2.50). In oceanic island environment the slopes of the islands are constantly exposed to humid airflow and intensive rainfall events (Fig. 2.50). Also, the common volcanic origin of the oceanic islands predetermines the major geometrical form of an island with steep to gentle slopes as it can be seen perfectly on the slope analysis map of Samoa. The regular slope angle values that are near to the angle of repose where loose material naturally tend to slide are the regions where other geological factors together can create regions where relatively little trigger can cause landslide generation. Mafic volcanic rocks such as basalt that is the main rock type of Samoa rich in iron and magnesium and tend to go through alteration where iron-rich clay minerals can form, especially along lava flow unit boundaries, breccia horizons or along alternating pyroclastic successions where coarse fine-grained members forming complex stratigraphy (Fig. 2.51). Landslides are not directly associated with volcanism, but volcanism also can cause landslides, and such collapse events can lead to submarine debris avalanche formation as outlined in the previous section. Landslides however in a young volcanic island in tropical environment can be triggered due to volcanic-related processes, but it is also possible that climatic-eustatic sea-level during the LGM (Williams et al. 2014) may have played a role in influencing failure of the flank of volcanic islands such as it has been suspected in Tau Island in American Samoa (Williams et al. 2014). Landslides are a common morphological feature in various scales in any Pacific Island with significant morphology such as Samoa. They can form in the razorback hills (Fig. 2.49) or in the steep coastal environment (Fig. 2.50). Rockfall and landslide hazards are commonly associated with the Fagaloa and Salani Volcanic Formations of Samoa (Figs. 2.51 and 2.52). A combination of rockfall and landslide killed several people on northeast Upolu Island in 1981 based on mouth-to-mouth information from local people of Falevao village. Skowron (1987) not fully explained this event however landslide and rockfall had been driven by an

Fig. 2.33 Thin pahoehoe lava flow surrounding the London Missionary Society Church of Saleaula (13°27′6.25″S, 172°19′50.30″W)

earthquake of 7.5 magnitudes at a depth 31 km with an epicentre of 60 km to the southwest of Samoa (Skowron 1987). Based on data from the National Geophysical Data Center/World Data Service (NGDC/WDS) there was no extensive damage in infrastructure at any other parts of Samoa. On 25th of June1917 an earthquake of 8.5 magnitudes lasted for 2 min generated a landslide and associated tsunami (12 m) on both main islands (Upolu and Savai'i) however, no loss of life was reported (Skowron 1987). The epicentre of this earthquake located about 15.50°S 173.00°W at a depth of 10 km. It described that the ground was agitated like the waves of the ocean generated many subsidence parts (up to 3 m^2 or more), massive landslides together with many acres of trees were uprooted (https://paperspast.natlib.govt.nz/newspapers/THD19170717.2.34—accessed on 19.11.2024).

2.5.3 Earthquakes

Earthquakes are significant geohazards associated mostly with regions along plate margin. As Samoa is in the proximity to the major trench along the Pacific Palte subducting beneath its older segment, the island is exposed tom potential high magnitude subduction earthquakes despite the island itself is already within the Pacific Plate. Earthquakes in geoheritage perspective are rarely explored partially because they are rather associated with dark geocultural events as a consequence of massive destructions (Chiang et al. 2021). There are only a handful detailed study exists where the link between earthquakes and their potential geoheritage value is articulated (Cunningham 2021; Das et al. 2024; Fort et al. 2018; Henriques 2023; Novković et al. 2022; Pacheco and

Fig. 2.34 Pahoehoe lava flow that entered the London Missionary Society Church of Saleaula (13°27′6.25″S, 172°19′50.30″W)

Cachão 2021; Song et al. 2023; Wang 2015; Wang et al. 2019; Wang and Tian 2014). Earthquake also commonly associated with volcanism and the volcanic geoheritage is interconnected with the earthquake geoheritage (Zhang et al. 2022). Earthquakes are commonly mentioned in the context of tectonic or structural geology geoheritage but also dedicated research that are directly addressing the tectonic geoheritage values are rare (Filocamo et al. 2019; Frassi et al. 2022; Ibanez et al. 2021; Maurya et al. 2023; van Wyk de Vries et al. 2018). In Samoa, while earthquakes are common, their direct effect in form of resulting landforms, landscapes or damaged geocultural elements are rare to see. Information on the most recent high intensity earthquakes is also scattered in various databases mostly outside of Samoa.

The 1917 Samoa earthquake affected the Samoan Islands, on June 26, 1917, at 18:49. The epicentre of the magnitude 8.0 earthquake was in the southwest of the Samoan Islands (Fig. 2.53). It is the worst earthquake in Samoa in 50 years. Though there were no casualties, residents on the island were frightened by the tsunami and sought refuge in higher ground (Sischka et al. 2022). The earthquake damaged two churches and several homes. A tsunami also flooded the islands and washed away homes (Sischka et al. 2022). At least two large earthquakes occurred since 1917; in second of April 1977 a magnitude 7.6, while recently on 29th September 2009 a magnitude 8.1 occurred. The effect of the 2009 September earthquake was significant due to the moderate tsunami it generated.

Fig. 2.35 Small lava spatter cone along the Falealupo Road in Seuseu area in western Savai'i is a perfect accessible location to demonstrate short-lived, small volume lava spatter dominated eruption site as an example for this type of volcanic hazard and keeping the cone as a good example for geoeducation purposes (13°31′17.50″S, 172°45′10.36″W)

Fig. 2.36 Small lava spatter cone along Falealupo Road in Western Savai'i exposing scoria, agglutinate and clastogenic lava in the exposed local quarry wall (13°31′17.50″S, 172°45′10.36″W)

Fig. 2.37 Google Earth Pro satellite 3D view toward north showing the 1902 eruption site as an example of lava spatter-fed small lava flow emission in the dorsal range of Savai'i. Note steep step-like escarpment toward the northern circum-island road (yellow line in the far view). Also note the more or less east west aligned spatter complex of Mount Mata o le Afi suggesting a lava fountain fed fissure eruption. Note Mount Mu deep pit crater because of lava drainage in the lower flank of the edifice

Fig. 2.38 Typical small lava spatter dominated cone in western Upolu (13°53′21.05″S, 171°55′53.21″W)

Table 2.4 Summary of eruption scenario of a short-term spatter-cone eruption derived from volcanic literature and historic reports of the 1902 event

Timeline	Events	Impacts/hazards	Actions required	Geoheritage/geoconservation
0–10 days	Earthquakes detected on seismographs in 1902—several were felt by people	Fear/unsettling of the population	Identification of eruption zone. Public warning Installation of volcanic monitoring network—seismic	There is no evident location to connect any precursory events (their results) to such type of eruptions
10–20 days	Eruption starts, producing small-scale explosions to build up one or more spatter cones	Vent area—burial of c. 200 m radius by spatter cone Volcanic bombs thrown out to 1.5 km radius—common in 1 km from vent Ashfall several cm thick in areas up to 3 km away	Evacuation of people in a 3 km radius of vent. Evaluation of lava flow pathways and potential new hazard areas Evaluation of wind direction and ash-fall hazards Public information Regular aerial monitoring of eruption	Mount o le Afi spatter cone complex is a good geosite but it is largely inaccessible
20–30 days	Lava flows begin—either from an opening of the crater (wind direction or collapse influenced) or from the base of the cone	Burial by lava of areas up to 3 km from source Minor ashfall	Evacuation of lava flow pathway and a down-slope projection from the pathway up to 10 km from the flow front Regular aerial monitoring to predict lava paths. Regular public information updates	Small, commonly spatter-fed lava flows that relate to the cones are the best geosites to document such geological processes such as those of the Mount o le Afi
1–3 months	Variable levels of activity, small flows from other sides of the cone, cone growth and inward collapse, small-scale explosions, finishing suddenly or gradually winding down over several months	Burial by lava of minor new areas minor ashfall	Maintenance of evacuation areas. Regular aerial monitoring Wind/weather monitoring Regular public information updates When eruptions have ceased for >6 months, wind down operations and evacuation restrictions	Small pit craters associated with a restricted row of cones mostly in the dorsal ridges of both islands. In Western Upolu there are some vegetated but accessible collapsed pit craters (Fig. 2.38)

Fig. 2.39 Phreatomagmatic pyroclastic rocks collapsing due to undercut erosion by the incoming waves in the SW side of Tafua Savai'i peninsula in south Savai'i (13°47′49.66″S, 172°15′2.42″W)

Fig. 2.40 Apolima Island between the two main Samoan islands is an eroded tuff cone formed through Surtseyan style hydrovolcanic processes (13°49′24.19″S, 172°9′6.46″W)

Fig. 2.41 Regular gully network in large, broad tuff ring in the eastern regions of Upolu resembling the presence of fine ash-dominated deposits likely formed due to magma and water explosive interaction (14°0′0.39″S, 171°30′43.30″W)

Table 2.5 Summary of eruption scenario for an explosive phreatomagmatic eruption (near-shore/offshore)

Timeline	Events	Impacts/hazards	Actions required	Geoheritage/geoconservation
0–10 days	Earthquakes detected on seismographs—some felt by people nearby 1 or 2 days before eruption	Fear/unsettling of the population	Identification of eruption zone. Public warning Installation of volcanic monitoring network—e.g., seismic	No geosites where the result of precursory events can be traced
10–12 days	Eruption starts, often in underwater areas (e.g., lagoons or near shore). Explosions begin immediately to penetrate water surface, producing vertical steam-rich ash columns up to 2 km high and laterally moving pyroclastic surges. Most eruption occurs underwater, and vent is drowned	Volcanic bombs thrown out to 500 m radius. Pyroclastic surges moving at >30 m/s laterally out up to 500 m from the vent Ashfall several cm thick in areas up to 3 km away Small seiche-like tsunami waves formed as eruption breaks surface and as surges are generated, affecting areas up to 2 km from source Volcanic gas/vog may form	Immediate evacuation of people in a 3 km radius of vent Evaluation of wind direction and ash-fall hazards Public information updates. Regular aerial monitoring of eruption	Basal phreatomagmatic successions with accidental lithic-rich horizons, even breccia horizons such as some coral fragment rich beds in the basal sections of Namua or Fanuatapu Islands
12–20 days	Island of volcanic debris is constructed around the vent area, reducing access of water—since some water still accesses the vent—eruption intensity increases with larger surges and more ash-rich columns	Volcanic bombs thrown out to 2 km radius. Pyroclastic surges moving laterally at >30 m/s laterally out up to a 2 km radius, structures and crops in their paths destroyed Ashfall several cm thick in areas up to 5 km away—minor ashfall can occur up to 10 km away Small tsunamis continue	Expansion of evacuation zone to a 5 km radius Regular aerial monitoring of eruption Monitoring of wind direction/strength to provide ashfall forecasts Regular public information updates	Main successions of the phreatomagmatic sequences along the shoreline of Tafua Savai'i where glassy pyroclast-rich lapilli tuffs are exposed. These rocks are massive due to their proximal location and form dark beds that are subhorizontally bedded. Occasional undulating beds with low angle crossbedding suggest their horizontal transportation mode

(continued)

Table 2.5 (continued)

Timeline	Events	Impacts/hazards	Actions required	Geoheritage/geoconservation
20 days to 3 months	Activity wanes as magma supply exhausted or when island fully formed around vent to block water access. In latter case, eruption intensity decreases, and spatter or lava flow activity begins	As above, with decreasing areas affected. Onset of spatter/lava activity coincides with cessation of surges and significant ashfall Lava flow onset will begin to cover nearby areas, interaction with seawater may produce vog	Maintenance of evacuation areas. Regular aerial monitoring Wind/weather monitoring. Regular public information updates When eruptions have ceased for >6 months, wind-down operations and evacuation restrictions	The scoria cone complex over the Tafua Savai'i peninsula is the best candidate to show that phreatomagmatic phase can reach pure magmatic explosive or even effusive stages as the Tafua Savai'i peninsula shares very similar architecture as Surtsey indicating that the original edifice emerged over the sea surface and due to the gradual cut of having water access to the hot magma switched the eruption to magmatic volatile-driven explosive and subsequently effusive eruptions

Fig. 2.42 Small spatter-scoria cones in eastern Savai'i (13°34′8.89″S, 172°17′20.67″W)

Fig. 2.43 Small scoria cone with spatter in its proximal edifice in eastern Savai'i (13°34′8.89″S, 172°17′20.67″W)

Fig. 2.44 Highly vesicular scoria with irizating blue colour from the small scoria cone (13°34′8.89″S, 172°17′20.67″W) in eastern Savai'i. The origin of such bright blue colour is unclear but the most common explanation that it is just a rapidly chilled outer glassy zone of a young eruptive products suggesting that the scoria cone must be very young (potentially Holocene) and the ash and lapilli pyroclasts cooled down very quickly. Recent study also suggested that it can be the result of a nano-scale layer in the vesicles (Debut et al. 2021) but also could be the result of chemical variations due to high temperature in the sudden chilled magma or the result of exhalation precipitating film layers of copper and arsenic minerals (Pekov et al. 2017). New studies also revealed that blue colours on glassy material can be the result of electrostatic discharge such as lighting (Feng et al. 2019). As electrostatic action is common in an ash-charged eruption cloud such origin is feasible in volcanic realm

While earthquakes are frequent in Samoa, there are little geological features remaining to see their geoheritage and certainly this is a subject for future research.

2.5.4 Tsunami

Certainly, many coastal villages that are often <10 m above sea level would be destroyed in case of significant tsunami generated by nearby subduction earthquakes. We currently have no idea of the frequency of such events in the Samoa area, and none have occurred from any of the world's oceanic islands in recent history. From geoheritage perspective, only the northern escarpment of Savaii as a major structural element can be noted and inventoried as a geosite however it is not so characteristic feature from the ground. Its scale and geometry can only be appreciated from various air photo or satellite imagery services. The most recent tsunami event in Samoa took place on 29th September 2009, caused by a M8.1 subduction earthquake in 15 km depth, just 190 km south of the Samoan Islands. Within 2 min of the earthquake rupture, two large magnitude 7.8 earthquakes occurred on the subduction zone interface. The islands were hit by maximum 14-m-high waves that inundated the coastal areas in places up to 700 m inland causing devastation (Fig. 2.54). The total of 183 people lost their life in this catastrophic event, and many has been displaced. Large fraction of the population moved inland and higher grounds after the tsunami and never returned since then to the coast. However, as the event fading from the memory of communities and the economic needs for

Fig. 2.45 Peridotite lherzolite nodules from a small scoria cone in eastern Savai'i (13°34′8.89″S, 172°17′20.67″W)

Fig. 2.46 Thin (less than a meter) and short lava flow preserved in the flank of a small scoria/spatter cone in western Savai'i (13°31′17.50″S, 172°45′10.36″W)

Fig. 2.47 Pit crater wall of the Tafua Savai'i cone (13°47′10.72″S, 172°15′9.31″W)

Fig. 2.48 Tafua Savai'i complex cone that is sitting over a phreatomagmatic tuff ring (13°47′10.72″S, 172°15′9.31″W)

Table 2.6 Summary of eruption scenario for an explosive scoria eruption

Timeline	Events	Impacts/hazards	Actions required	Geoheritage/geoconservation
0–10 days	Earthquakes detected on seismographs—some felt by people nearby, 1 or 2 days before eruption	Fear/unsettling of the population	Identification of eruption zone. Public warning Installation of volcanic monitoring network—seismic	No direct geosites that preserved precursory event-related features
10–12 days	Eruption starts, with explosive blasts and initial ejection of soils and country rocks from the newly formed vent area	Volcanic bombs thrown out to 500 m radius Possible pyroclastic surges moving laterally at >30 m/s out up to 500 m from the vent if eruption occurs in a wet area. Ashfall several cm thick in areas up to 3 km away	Immediate evacuation of people in a 3-km radius of vent Evaluation of wind direction and ashfall hazards Public information updates Regular aerial monitoring of eruption	Basal pyroclastic successions of scoria cones are not exposed hence no direct setting where the initial stage of a scoria cone formation has been recorded and the resulted geological feature preserved
12–20 days	A scoria cone or chain of cones is rapidly formed by frequent ongoing explosive bursts. A 5–10 km high column of ash/lapilli is formed	Volcanic bombs thrown out to 2 km radius cone formed of up to c. 500 m diameter, burying all surroundings Ashfall several cm thick in areas up to 10 km away—minor ashfall can occur at greater distances	Expansion of evacuation zone to a 5 km radius and perhaps further in down-wind areas if ashfall is strong Regular aerial monitoring of eruption. Monitoring of wind direction/strength to provide ashfall forecasts Regular public information updates	Geosites of scoria cones along the western Savai'i road are also good examples for small scoria cone forming events. The most accessible location that is a perfect geosite is along the inland road of eastern Savai'i, near Vaiola (13°34′8.70″S, 172°17′21.44″W) where a group of small spatter-scoria cones located (Fig. 2.38), exposing their pyroclastic successions (Figs. 2.39 and 2.40) that is also abundant in mantle-derived peridotite lherzolite nodules, a rare site to have them in Samoa (Fig. 2.41)
20 days to several years	Activity wanes and waxes as magma supply changes—there may be a transition to lava flow activity as later-erupted magma often has progressively lower gas-content	As above with varying areas affected depending on wind direction changes. Onset of spatter/lava activity coincides with cessation of significant ashfall Lava-flow onset will begin to cover nearby areas	Maintenance of evacuation areas. Regular aerial monitoring Wind/weather monitoring Regular public information updates. When eruptions have ceased for >6 months, wind down operations and evacuation restrictions	Gradually thinning scoriaceous pyroclastic successions in the top section of exposed scoria cone edifices where such geological features have likely occurred. Only in some locations like the quarried scoria cones in the Vaiola region where the exposed quarry walls reaching the upper section and some textural changes visible preserved

Fig. 2.49 Landslide scars in the old part of NE Upolu (13°57′10.88″S, 171°36′33.73″W)

tourism, significant rebuilding took place and small tourism ventures gradually returning to the disaster hit zones. The 2009 tsunami and earthquake highlighted the need for focused research to understand tsunamis. Also, scientific research and community based participatory methods resulted several good outcomes to design potential evacuation routes and to do list in case of future tsunamis (Fig. 2.55). Intensified research followed the 2009 tsunami revealed the changes on the coastal regions (Fig. 2.56). New sites appeared that show clearly the impact of the tsunami and they could be perfect geosites for tsunami hazard resilience programs through their geoheritage values. Such research and associated implementation strategies are subject for future work to develop, co-develop a tsunami hazard resilient society in Samoa.

Several reconnaissance mapping and site visit revealed that there is substantial number of locations where signs of former tsunami inundation can be reconstructed including locations that are already part of geotouristic developments such as the southern coastal walkway of the lava flow fields of O Le Pupu-Pu'e National Park in southern Upolu. The coastal walk runs across the young lava field reached the ocean and formed fantastic cliffs that are about ~10 m above the current sea level (Fig. 2.57). In the walking track there are large number of locations where marine shells beds with pebbles are common as suspected tsunami deposits (Fig. 2.58). Such locations clearly require further investigation to confirm their origin then could form significant Geosites that could be used for geohazard education.

Fig. 2.50 Debris fans are common, even covered by vegetation, along most of the road cuts of the around the-island roads (13°53′0.92″S, 171°37′7.00″W)

Fig. 2.51 Large landslide occurred on 12th February 2018 along the SE coast of Upolu, temporarily cutting the coastal road off (14°2′40.52″S, 171°27′47.51″W)

Fig. 2.52 Local landslide along a steep gorge near to Togitogiga waterfall exposing a complex volcaniclastic succession (14°0′52.83″S, 171°43′5.99″W)

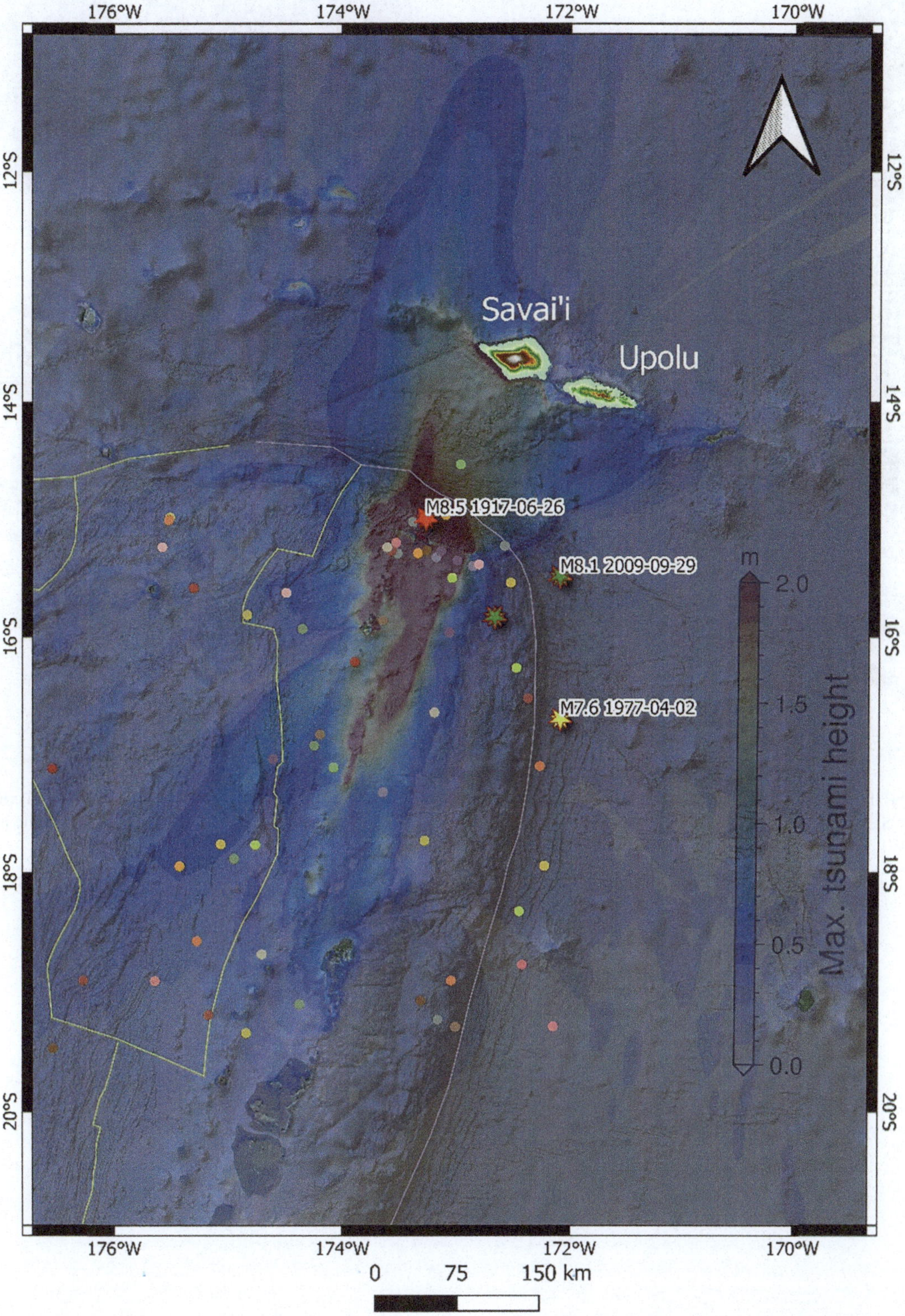

Fig. 2.53 Large magnitude earthquake epicentres nearby Samoa in the last 100 years marking the three largest events from what the 1917 and 2009 generated significant tsunami. The map shows the larger than M6.5 earthquake locations recorded since 1900 (USGS Earthquake catalogue). A modelled tsunami wave propagation (Sischka et al. 2022) is shown over the Google Earth Pro seafloor map and the DEM of Samoa Islands. Note that larger than 2 m wave height reached mostly the southern coast of Savai'i. In 2009, an estimated 14 m high tsunami wave hit the SE coast of Upolu. Note the double green star marking the pair of the 2009 earthquake locations. The tsunami wave propagation in the 2009 event was more north-eastward, and hit directly Upolu (Bosserelle et al. 2020)

Fig. 2.54 Only the foundation left of many of the light structure houses in the SE coast of Upolu near Saleapaga (14°2′23.78″S, 171°29′44.07″W). The remaining houses could serve good geoheritage memorials however the dark disaster aspect of them commonly lead for complete demolish or rebuilding of them to erase the negative memory within the communities

Fig. 2.55 Tsunami warning sign at Savaia Village (13°57′17.18″S, 171°57′33.60″W) marking the evacuation routes in case of future tsunami ()

Fig. 2.56 New lagoons developed after the 2009 tsunami event in eastern Upolu because of the coastal morphology changes. These sites can be utilized as significant geoheritage sites to demonstrate the power and impact of tsunamis in the coastal systems (14°1′4.78″S, 171°25′27.09″W)

Fig. 2.57 Coastal lava cliffs in southern Upolu forming about 10 m high wall that is covered by suspected tsunamigenic deposits (14°2′42.52″S, 171°44′44.67″W). View looking toward west along the southern coast of Upolu

Fig. 2.58 Shell and coral beds about 10 m above the current sea level in southern Upolu (14°2′40.72″S, 171°44′50.21″W)

2.6 Coastal Architecture

In Samoa three main type of coastal setting has been distinguished (Richmond 1992); (1) wide fringing reef transitional to shallow barrier reef, (2) cliffed coast with little or no reef development, and (3) fringing reefs and narrow coastal strip or beaches, barrier spits, coastal swamps associated with streams (Fig. 2.59).

The coastal geodiversity of Samoa is important as the coastal regions are the prime destinations for tourist activities. Prime beach sections are those located in the SE coastal regions of Upolu where the most advanced tourism developments to utilize the calm coastal waters and white coral sands forming the main attraction of the region (Fig. 2.60).

Sandy beaches are also common behind small coral reef blocked regions where main valleys run to the coastal areas such as those in the NW coastal section of Upolu (Fig. 2.61). In those regions the narrow lagoons provide shelter to a shallow marine coastal ecosystem and provide attractive backdrop for the cliffed shorelines (Fig. 2.61).

These narrow coastal strips are somehow in contrast with the wide fringing reef regions, essentially forming the main coastal landscape in the most populous NW and N region of Upolu (Fig. 2.62). In these areas, especially in the region of Manono shield volcanic island the shallow marine region provides a calm and harmonic landscape where local communities are sustaining their life from reef fishing. This region also suspected to be main archaeological sites in the early Polynesian history when sea level was lower than today, and the coastal regions formed a coastal plain capable to host larger human communities (Kane et al. 2017; Petchey 2001).

Rocky, cliffed coastal areas are located mostly in those areas where young lava flow fields reached the coastal areas. Wave cut erosion and the extensive, tube-fed lava flow network generated a strong coastal interface that is mechanically abraded due to the wave actions (Fig. 2.63). Extensive steep coasts along the Holocene O le Pupu Pue lava flow fields in southern Upolu provides a unique geological environment upon diverse coastal habitat developed (Fig. 2.63). Along the coast, only narrow beach formed that are littered by large semi-rounded boulders. The coastal section is nearly inaccessible, but from side views the architecture of the lava fields can be seen clearly (Fig. 2.63). Along the rocky shorelines rock arches (Fig. 2.64) and separated islets (Fig. 2.65) can be observed because of the high energy coastal environment.

Rocky, cliffed coastal areas are the main coast types in Savai'i. Various types of rocky cliffed coast can be seen where distal or more proximal lava fields forming the land—ocean interface. The most western region of Savaii is typical of relatively thin (<10 m) cliffs of a lava fields (Fig. 2.66). Similarly, in the entire south coast of Savaii high cliffed coasts are the most common. In these regions waterfalls are commonly develop right on the narrow beach coves where surface water can follow the erosional units of former lava flows, generating typical step like morphology until they reach the ocean. In those areas

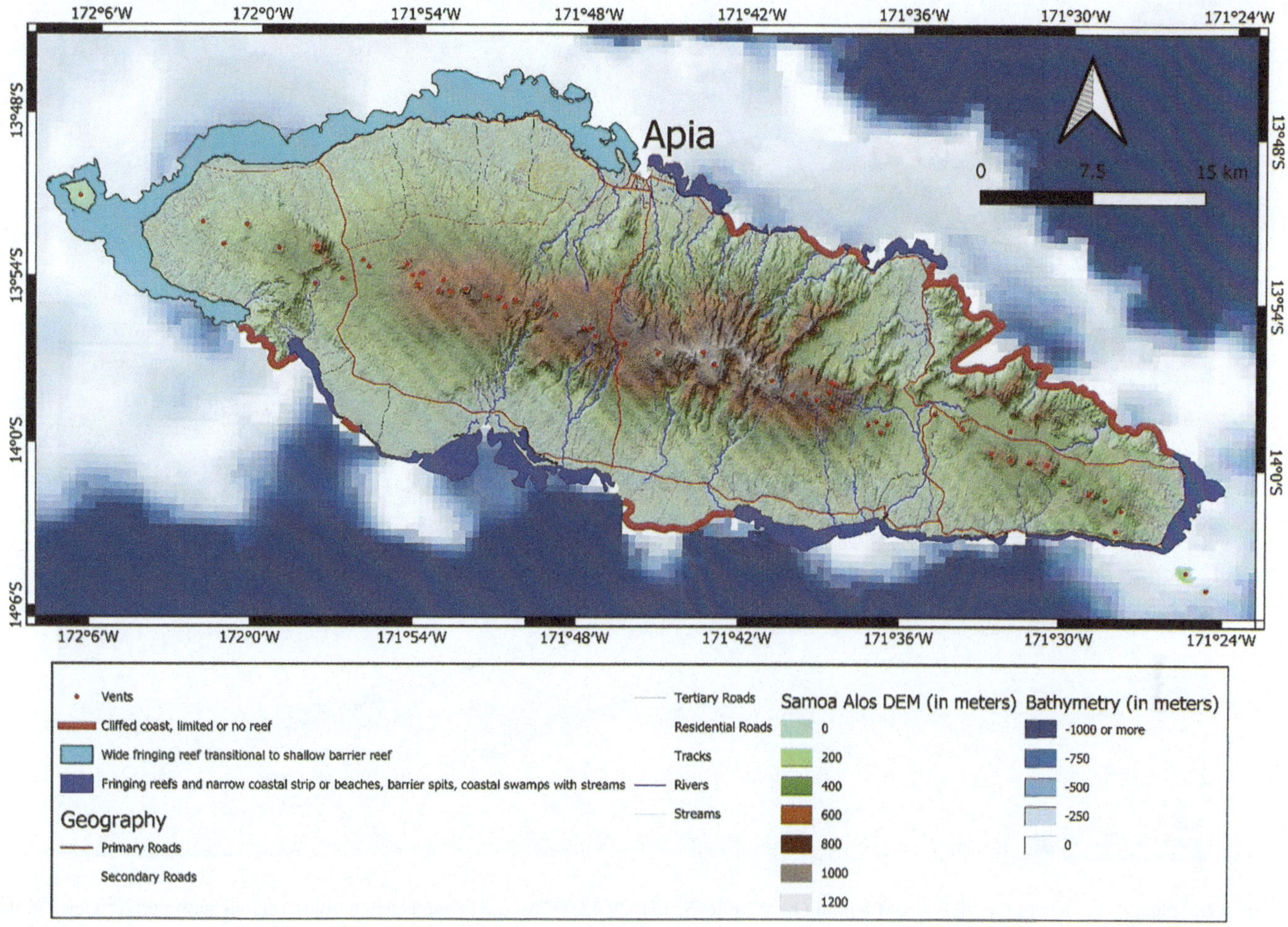

Fig. 2.59 Coastal types of Upolu (Richmond 1992) based on the presence and type of reefs recognized. The map showing the ALOS PALSAR DEM based topography as well as a coarse bathymetry of the region. As Upolu hosts the oldest rocks of the Samoan Islands reef formation is common due to the subsidence of the landmass creating extensive reef network around the island. In contrast Savai'i coastal regions has only limited reef development in their SE and NE sections while rocky, cliffed coast are common

where medial tube fed lava flows cut by the coastal erosion; blow holes can form. The wave action can push water into the tubes by the incoming waves (especially in the time high tide arrives), pressing water out in the former skylines of the lava tubes, generating spectacular blow holes such as those in the Alofaaga coastal section in Southern Savai'i (Fig. 2.67).

As Savai'i Island consists of young volcanic rocks, dominantly derived from basaltic volcanic eruptions, its coastal areas are commonly the playground where distal lavas entered to the ocean. In most cases the present-day shoreline is the results of the active wave cut erosion and the associated gradual retreat and collapse of the shoreline. During the process of lava and ocean interaction, abundant fragmented material can form producing large volume of clasts to be carried away from the interaction site along the coast. At the same time even rootless volcanic cones (e.g., littoral cones) can form that are gradually fed by magma through tube network and fragmented by the magma and water interaction along the coast (Fisher 1968; Green 1984;

Fig. 2.60 Sandy beaches in the SE coast of Upolu (14°2′24.86″S, 171°29′42.95″W) are prime beach activity regions with perfect views to the nearby offshore islands. Note that however this region was heavily impacted by the 2009 tsunami and rebuilding is still ongoing process

Fig. 2.61 Typical fringing reef blocked short coastal lagoon area in SW Upolu, closed by cliffed shorelines to the west (13°57′15.57″S, 171°57′34.42″W)

Fig. 2.62 Wide fringing reef region in the western shallow marine lagune system in offshore Upolu (13°53′26.25″S, 172°3′58.62″W). In the background Manono shield volcano and Apolima tuff cone top is visible

Fig. 2.63 Coastal section in southern Upolu exposes complex Holocene lava flow fields (14°2′38.26″S, 171°44′57.23″W)

Holt et al. 2021). The production of large volume of fragmented volcanic debris can generate a rapid build-up of sand spits further away the interaction site along the coast. As the lava flows also entering to the ocean along longitudinal network of flow lobes, the coastal complexity of the interface ca generate regions where nearby debris can form barriers and local lagoons. In the 1905–1911 Matavanu eruption such barriers created lagoons that gradually occupied over time mangrove forests (Fig. 2.68). Further away typical volcaniclastic fragment-dominated spit (Fig. 2.69) formed facilitating the development of coastal lagoons in northern Savai'i.

Particularly in Savai'i Island, along the coastal regions young lava flows that were not as voluminous forming well distinguished zones that separate coastal section or forming barriers to alter the surface run off. As a result, in those regions where alluvial systems reach the coastal plain, complex coastline can form where the freshwater addition can facilitate to form mangrove forests and complex coastal habitat (Fig. 2.70). Somehow similar scenario can be seen in southern Upolu as well (Fig. 2.71). In the southern regions of Upolu where large capacity fluvial networks capable to carry large volume of terrigenous detritus the coastal lagoons form typical mixed water environment, where mud dominated sedimentation dominates (Fig. 2.71).

Fig. 2.64 Rock arch developed in the southern Upolu coastal region (14°2′41.30″S, 171°44′46.63″W). Note the reddish and black colours marking brecciated and coherent basaltic lava body of a complex lava flow entered to the ocean here

Fig. 2.65 Separated islets along the southern shoreline of Upolu (14°2′42.76″S, 171°44′49.90″W). The main arch of the islets is a remnant of an exhumed lava tube reflecting the size of the lava tube network fed this Holocene volcanic system

Fig. 2.66 Pahoehoe lava flow field forming rocky cliffed coast in at the Fagalele Bay area in western Savai'i (13°29′41.56″S, 172°46′13.30″W). Note the fresh pahoehoe lava surface textures in the foreground and the shield like topography of Savaii in the distance

Fig. 2.67 Alofaaga Blowholes are spectacular examples of the lava tube and tidal interaction (13°48′5.99″S, 172°31′18.65″W)

Fig. 2.68 Complex lagoon system in the western edge of the 1905–1911 Matavanu lava flow field (13°27′1.74″S, 172°19′48.32″W)

Fig. 2.69 Volcaniclastic debris spit that formed by remobilized volcanic fragments from the interaction zone of the 1905–1911 Matavanu lava flow and the ocean (13°26′51.12″S, 172°20′14.80″W)

Fig. 2.70 Thin lava flow lobes forming complex coastal regions where lagoons with mangrove can form while fresh water ponded (13°28′13.20″S, 172°25′28.18″W)

Fig. 2.71 Typical coastal area in southern Upolu where the terrigenous sediment input dominates the lagunes to form muddy coastal regions (13°59′47.64″S, 171°49′47.43″W)

2.7 Conclusion

Samoa geological history is dominated by volcanic processes governed by the SW Pacific plate tectonic setting. The islands were dominated by volcanism governed by hot spot volcanism that formed characteristic oceanic island basalts. While the geological history clearly showing that the evolution of the island is not simple the main framework of hot spot volcanism is generally accepted. The resulting geological formations therefore dominated by basaltic rock types, mostly coherent lava flows forming broad lava shields determining the main geological features. Both islands experienced rejuvenated, relatively small volume and restricted volcanism, that covered most of the early shields by small to medium size scoria cones and extensive lava fields. The youngest eruption on the island took place in 1906 lasting few years until 1911 leaving behind a distinct young lava field in the NE of Savai'i. Mostly Upolu shows clear morphological evidence of an erosional stage followed the shield building phase that fits well to the general model of ocean island volcanism. The geological architecture of the islands strongly influencing the geomorphology of Samoa. The broad shields and the deep incised canyons are typical geomorphological features. While along the oldest part of the island the erosion cut deeply into the older volcanic successions and generating great morphodiversity, the general trend of the island's morphology is rather settled and dominated by the underlying shield morphology. The geology and geomorphology together will provide the fundamental basis of geodiversity estimates outlined in the next chapter.

References

Addison DJ, Matisoo-Smith E (2010) Rethinking Polynesians origins: a West-Polynesia triple-I model. Archaeol Ocean 45(1):1–12

Agustín-Flores J, Németh K, Cronin SJ, Lindsay JM, Kereszturi G, Brand BD, Smith IEM (2014) Phreatomagmatic eruptions through unconsolidated coastal plain sequences, Maungataketake, Auckland Volcanic Field (New Zealand). J Volcanol Geotherm Res 276:46–63

Anderson T (1910) The volcano of Matavanu in Savaii. Q J Geol Soc 66(1–4):621–639

Avis C, Montenegro A, Weaver A (2007) The discovery of western oceania: a new perspective. J Island Coast Archaeol 2(2):197–209

Bachèlery P, Villeneuve N (2013) Hot spots and large igneous provinces. In: Treatise on geomorphology, vol 1–14. Elsevier, Amsterdam, pp 193–233

Becerril L, Larrea P, Salinas S, Mossoux S, Ferrés D, Widom E, Siebe C, Martí J (2021) The historical case of Paricutin volcano (Michoacán, México): challenges of simulating lava flows on a gentle slope during a long-lasting eruption. Nat Hazards 107(1):809–829

Biain A, León R, Urgeles R, Somoza L, Medialdea T, Ferrer M, Gonzalez FJ (2016) Onshore and offshore geomorphological features of the El Golfo debris avalanche (El Hierro, Canary Islands). In: Advances in natural and technological hazards research. Springer, Berlin, pp 83–92

Bird P (2003) An updated digital model of plate boundaries. Geochem Geophys Geosyst 4(3):1027

Birkeland C, Craig P, Fenner D, Smith L, Kiene WE, Riegl BM (2008) Geologic setting and ecological functioning of coral reefs in American Samoa. In: Coral reefs of the world. Springer, Berlin, pp 741–765

Boschman LM, van Hinsbergen DJJ (2016) On the enigmatic birth of the Pacific Plate within the Panthalassa Ocean. Sci Adv 2(7):e1600022

Bosserelle C, Williams S, Cheung KF, Lay T, Yamazaki Y, Simi T, Roeber V, Lane E, Paulik R, Simanu L (2020) Effects of source faulting and fringing reefs on the 2009 South Pacific Tsunami inundation in southeast Upolu, Samoa. J Geophys Res Oceans 125(12):e2020JC016537

Boulesteix T, Hildenbrand A, Soler V, Quidelleur X, Gillot PY (2013) Coeval giant landslides in the Canary Islands: implications for global, regional and local triggers of giant flank collapses on oceanic volcanoes. J Volcanol Geotherm Res 257:90–98

Calvari S, Pinkerton H (1999) Lava tube morphology on Etna and evidence for lava flow emplacement mechanisms. J Volcanol Geotherm Res 90(3):263–280

Calvari S, Giudice G, Maugeri R, Messina D, Morgavi D, Miraglia L, La Spina A, Spampinato L (2024) Complex lava tube networks developed within the 1792–93 lava flow field on Mount Etna (Italy): insights for hazard assessment. Front Earth Sci 12

Carracedo JC (1999) Growth, structure, instability and collapse of Canarian volcanoes and comparisons with Hawaiian volcanoes. J Volcanol Geotherm Res 94(1–4):1–19

Carracedo JC (2014) The Teide volcano, Tenerife, Canary Islands. In: World geomorphological landscapes. Springer, Berlin, pp 257–272

Cas R, Wright JV, Giordano G (2024) Basaltic and other low-viscosity lavas. In: Cas R, Giordano G, Wright JV (eds) Volcanology: processes, deposits, geology and resources. Springer International Publishing, Cham, pp 229–309

Chiang CS, Chu TM, Chou WH, Lee SH, Wang JF (2021) Challenges in the preservation of disaster remains—example of the Chelungpu fault preservation park. J Disaster Res 16(2):201–209

Cibik T (1999) Petrology of pre-shield, shield and post-erosional lavas of Western Samoa, SW Pacific. The HKU (Univeristy of Hong Kong) Scolars Hub, MSc thesis, p 196

Clark JT (1993) Radiocarbon dates from American Samoa. Radiocarbon 35(2):323–330

Clark JT, Michlovic MG (1996) An early settlement in the Polynesian homeland: excavations at'Aoa Valley, Tutuila Island, American Samoa. J Field Archaeol 23(2):151–167

Clouard V, Bonneville A (2002) How many Pacific hotspots are fed by deep-mantle plumes? Geology 29(8):695–698

Coello-Bravo JJ, Márquez Á, Herrera R, Huertas MJ, Ancochea E (2020) Multiple related flank collapses on volcanic oceanic islands: evidence from the debris avalanche deposits in the Orotava Valley water galleries (Tenerife, Canary Islands). J Volcanol Geotherm Res 401:106980

Cole PD, Guest JE, Duncan AM, Pacheco JM (2001) Capelinhos 1957-1958, Faial, Azores: deposits formed by an emergent Surtseyan eruption. Bull Volcanol 63(2–3):204–220

Couling M (2014) Tsunami risk perception and preparedness on the east coast of New Zealand during the 2009 Samoan Tsunami warning. Nat Hazards 71(1):973–986

Cronin SJ, Taylor PW, Malele F (2000) Final report—Savai'i, volcanic hazards project Samoa,October 2000. SOPAC technical report (Suva, FIji), vol 343, pp 1–34

Cronin SJ, Bebbington M, Lai C (2001) A probabilistic assessment of eruption recurrence on Taveuni volcano, Fiji. Bull Volcanol 63(4):274–288

Cronin, S.J, Neall, VE (2001) Holocene volcanic geology, volcanic hazard and risk on Taveuni, Fiji. New Zealand Journal of Geology and Geophysics 44:417–437

Cronin SJ, Bonte-Grapentin M, Németh K (2006) Samoa technical report—review of volcanic hazard maps for Savai'i and Upolu. EU-SOPAC project report, vol 59, pp 1–27

Cunningham D (2021) The case for a globally recognized Geopark in the NE Gobi Altai region of Mongolia. Geoheritage 13(4):1–20

Dalrymple GB, Clague DA, Lanphere MA (1977) Revised age for midway volcano, Hawaiian volcanic chain. Earth Planet Sci Lett 37(1):107–116

Das A, Nauriyal KC, Prizomwala SP, Borgohain C, Pandey UR, Vedpathak C, Kouli DR, Deori N, Hassan C (2024) Geoheritage of Charaideo Moidams (Assam, India): a tale of human resilience and man-environment relationship. Geoheritage 16(3)

Davies H, Bani P, Black P, Smith I, Garaebiti E, Rodda P (2004) Oceania (including Fiji, Png and Solomons). In: Encyclopedia of geology. Elsevier, Amsterdam, pp 109–123

Debut A, Toulkeridis T, Vaca AV, Arroyo CR (2021) Origin of color variations of thin, nano-sized layers of volcanic cinder from the Sierra Negra Volcano of the Galapagos Islands. Uniciencia 35(2):1–13

Dickinson WR (1998) Geomorphology and geodynamics of the Cook-Austral Island-seamount chain in the South Pacific Ocean: implications for hotspots and plumes. Int Geol Rev 40(12):1039–1075

Duncan RA (1990) The volcanic record of the reunion hotspot. Proc, scientific results, ODP, Leg 115, Mascarene Plateau, pp 3–10

Esson J, Flower MFJ, Strong DF, Upton BGJ, Wadsworth WJ (1970) Geology of the Comores archipelago, Western Indian Ocean. Geol Mag 107(6):549–557

Feng T, Lang C, Pasek MA (2019) The origin of blue coloration in a fulgurite from Marquette, Michigan. Lithos 342–343:288–294

Fepuleai A (2016) Establishing a Holocene tephrochronology for Western Samoa and its implication for the re-evaluation of volcanic hazards. In: Unpublished PhD Thesis, University of the South Pacific, Suva, Fiji [Call No.: Pac QE 527.56.F47 2016; BRN: 1208019]; p 370, Suva

Fepuleai A, Németh K (2019) Volcanic geoheritage of landslides and rockfalls on a tropical ocean island (Western Samoa, SW Pacific). Geoheritage 11(2):577–596

Fepuleai A, Weber E, Németh K, Muliaina T, Iese V (2017) Eruption styles of Samoan volcanoes represented in tattooing, language and cultural activities of the indigenous people. Geoheritage 9(3):395–411

Fepuleai A, Németh K, Muliaina T (2021) Geopark impact for the resilience of communities in Samoa. SW Pacific Geoheritage 13(3)

Filocamo F, Rosskopf CM, Amato V (2019) A contribution to the understanding of the Apennine landscapes: the potential role of Molise Geosites. Geoheritage 11(4):1667–1688

Fisher RV (1968) Puu Hou littoral cones, Hawaii. Geol Rundsch 57(3):837–864

Fort M, Adhikari BR, Rimal B (2018) Pokhara (central Nepal): a dramatic yet geomorphologically active environment versus a dynamic, rapidly developing city. In: Urban geomorphology: landforms and processes in cities. Elsevier, Amsterdam, pp 231–258

Foulger GR (2007) The "plate" model for the genesis of melting anomalies. In: Special paper of the Geological Society of America, vol 430. Geological Society of America, Boulder, pp 1–28

Franchello G, Annunziato A (2012) The Samoa tsunami of 29 September 2009 early warning and inundation assessment. Sci Tsunami Haz 31(1):19–60

Frassi C, Amorfini A, Bartelletti A, Ottria G (2022) Popularizing structural geology: exemplary structural geosites from the Apuan Alps UNESCO Global Geopark (Northern Apennines, Italy). Land 11(8):1282

Fryer P (1974) Petrology of some volcanic rocks from the northern Fiji plateau. Bull Geol Soc Am 85(11):1717–1720

Garcia MO, Weis D, Jicha BR, Ito G, Hanano D (2016) Petrology and geochronology of lavas from Ka'ula volcano: implications for rejuvenated volcanism of the Hawaiian mantle plume. Geochim Cosmochim Acta 185:278–301

Garcia MO, Norman MD, Jicha BR, Lynn KJ, Jiang P (2024) Reexamining the Honolulu Volcanics: Hawai'i's classic case of rejuvenation volcanism. J Petrol 65(9):egae093

García MS, Nieto-Torres A, Del Pozzo ALM (2024) Reconstruction of the paleo-relief of the Xitle volcano area (Mexico City) and its impact on the trajectory of its lava flows through probabilistic simulations. Geofis Int 63(4):1241–1263

Giordano G, Cas R, Wright JV (2024) Explosive eruption styles, columns, and pyroclastic fallout deposits. In: Cas R, Giordano G, Wright JV (eds) Volcanology: processes, deposits, geology and resources. Springer International Publishing, Cham, pp 519–691

Gokon H, Koshimura S, Imai K, Matsuoka M, Namegaya Y, Nishimura Y (2014) Developing fragility functions for the areas affected by the 2009 Samoa earthquake and tsunami. Nat Hazards Earth Syst Sci 14(12):3231–3241

González-Riancho P, Aliaga B, Hettiarachchi S, González M, Medina R (2015) A contribution to the selection of tsunami human vulnerability indicators: conclusions from tsunami impacts in Sri Lanka and Thailand (2004), Samoa (2009), Chile (2010) and Japan (2011). Nat Hazards Earth Syst Sci 15(7):1493–1514

Goodwin ID, Grossman EE (2003) Middle to late Holocene coastal evolution along the south coast of Upolu Island, Samoa. Mar Geol 202(1–2):1–16

Grant-Taylor TL, Rafter TA (1962) New Zealand radiocarbon age measurements—5. N Z J Geol Geophys 5(2):331–359

Green J (1984) Littoral cones. In: Schwartz M (ed) Beaches and coastal geology. Springer US, New York, pp 519–520

Groom S, Barfod DN, Millar I, Downes H (2023) The Cumbre Nueva collapse (La Palma, Canary Islands): new age determinations and evidence of an isotopic excursion. J Volcanol Geotherm Res 433:107708

Hakim WL, Ramayanti S, Park S, Ko B, Cheong DK, Lee CW (2022) Estimating the pre-historical volcanic eruption in the Hatangang River volcanic field: experimental and simulation study. Remote Sens 14(4):894

Harpp KS, Hall PS, Jackson MG (2014) Galápagos and Easter: a tale of two hotspots. In: The Galapagos: a natural laboratory for the earth sciences. Wiley, New York, pp 27–40

Hart SR, Coetzee M, Workman RK, Blusztajn J, Johnson KTM, Sinton JM, Steinberger B, Hawkins JW (2004) Genesis of the Western Samoa seamount province: age, geochemical fingerprint and tectonics. Earth Planet Sci Lett 227(1–2):37–56

Hawkins AB (1987) Samoa: significance of its geological setting. In: Memoir—Geological Society of China, vol 9. The Society, Taipei, pp 159–174

Henriques MH (2023) Broadening frontiers in geoconservation: the concept of intangible geoheritage represented by the 1755 Lisbon earthquake. Geoheritage 15(2):57

Hill PJ, Tiffin DL (1993) Geology, sediment patterns, and widespread deformation on the sea floor off Western Samoa revealed by wide-swath imagery. Geo-Mar Lett 13(2):116–125

Hofmann AW, Farnetani CG (2013) Two views of Hawaiian plume structure. Geochem Geophys Geosyst 14(12):5308–5322

Holt SJ, McPhie J, Carey RJ (2021) Apparently 'dry' littoral rootless cones in Hawai'i formed by sustained, "confined" mixing of lava and sea water. J Volcanol Geotherm Res 419:107352

Houghton B, White JDL, Van Eaton AR (2015) Chapter 30—Phreatomagmatic and related eruption styles. In: Sigurdsson H (ed) The encyclopedia of volcanoes, 2nd edn. Academic, Amsterdam, pp 537–552

Ibanez K, Garcia MGM, Mazoca CEM (2021) Tectonic geoheritage as records of Western Gondwana history: a study based on a geosite's potential in the Central Ribeira Belt, Southeastern Brazil. Geoheritage 13(1)

Idárraga-García J, García-Varón J, León H (2021) Submarine geomorphology, tectonic features and mass wasting processes in the archipelago of San Andres, Providencia and Santa Catalina (western Caribbean). Mar Geol 435:106458

Ilanko T, Pittari A, Briggs R, Németh K, Rosenberg M (2024) Eruption and emplacement processes of the Pleistocene Barriball Road Tuff Ring, South Auckland volcanic field, New Zealand. J Volcanol Geotherm Res 452:108130

Jackmond G, Fonoti D, Matavai Tautunu M (2018) Samoa's hidden past: lidar confirms inland settlement and suggests larger populations in pre-contact sāmoa. J Polynesian Soc 127(1):73–90

Jackmond G, Fonoti D, Meleisea M, Schoeffel P, Durling M, Tautunu MM, Sahib M (2022) UTU: Sāmoa archaeology and cultural heritage database. Archaeol Ocean 57(2):95–110

Jackson MG, Hart SR, Konter JG, Koppers AAP, Staudigel H, Kurz MD, Blusztajn J, Sinton JM (2010) Samoan hot spot track on a "hot spot highway": implications for mantle plumes and a deep Samoan mantle source. Geochem Geophys Geosyst 11(12)

Jackson MG, Finlayson VA, Steinberger B, Konrad K (2024) When a plateau suppresses a plume: disappearance of the Samoan Plume under the Ontong Java Plateau. AGU Adv 5(4)

Jasiewicz J, Stepinski TF (2013) Geomorphons—a pattern recognition approach to classification and mapping of landforms. Geomorphology 182:147–156

Jeon Y, Ki J, Southcott D (2024) Korean geoheritage: the volcanic landforms of the Jeju Island UNESCO Global Geopark. Episodes 47(2):295–310

Jicha BR, Garcia MO, Lormand C (2023) A possible sea-level fall trigger for the youngest rejuvenated volcanism in Hawai'i. Bull Geol Soc Am 135(9–10):2478–2485

Johnson KTM, Sinton JM, Price RC (1986) Petrology of seamounts northwest of Samoa and their relation to Samoan volcanism. Bull Volcanol 48(4):225–235

Kane HH, Fletcher CH, Cochrane EE, Mitrovica JX, Habel S, Barbee M (2017) Coastal plain stratigraphy records tectonic, environmental, and human habitability changes related to sea-level drawdown, Upolu, Samoa. Quat Res (United States) 87(2):246–257

Kear D, Wood BL (1959) The geology and hydrology of Western Samoa. N Z Geol Surv Bull 63:1–90

Keating BH (1991) Geology of the Samoan Islands. In: Keating Barbara H, Bolton Barrie R (eds) Geology and offshore mineral resources of the Central Pacific Basin. Circum-Pacific Council for Energy and Mineral Resources, Houston, pp 127–178

Keating BH (1992) The geology of the Samoan Islands. In: Keating BH, Bolton BR (eds) Geology and offshore mineral resources of the Central Pacific Basin. Springer, New York, pp 127–178

Keating BH, Helsley CE, Karogodina I (2000) Sonar studies of submarine mass wasting and volcanic structures off Savaii Island, Samoa. Pure Appl Geophys 157(6–8):1285–1313

Kereszturi G, Németh K, Cronin SJ, Procter J, Agustín-Flores J (2014) Influences on the variability of eruption sequences and style transitions in the Auckland Volcanic Field, New Zealand. J Volcanol Geotherm Res 286:101–115

Kim JR, Lin SY, Oh JW (2022) The survey of lava tube distribution in Jeju Island by multi-source data fusion. Remote Sens 14(3):443

Konrad K, Jackson M, Steinberger B, Koppers A, Balbas A, Finlayson V, Konter J, Price A (2024) Toroidal flow around the Tonga slab moved the Samoan plume during the Pliocene. Geology 52(3):176–180

Koppers AAP, Russell JA, Jackson MG, Konter J, Staudigel H, Hart SR (2008) Samoa reinstated as a primary hotspot trail. Geology 36(6):435–438

Koppers AAP, Russell JA, Roberts J, Jackson MG, Konter JG, Wright DJ, Staudigel H, Hart SR (2011) Age systematics of two young en echelon Samoan volcanic trails. Geochem Geophys Geosyst 12(7)

Kwasnitschka T, Hansteen TH, Ramalho RS, Devey CW, Klügel A, Samrock LK, Wartho JA (2024) Geomorphology and age constraints of seamounts in the Cabo Verde archipelago, and their relationship to Island ages and geodynamic evolution. Geochem Geophys Geosyst 25(3):e2023GC011071

Lara LE, Díaz-Naveas J, Reyes J, Jicha B, Orozco G, Tassara A, Kay S (2018) Unraveling short-lived rejuvenated volcanism and a rapid transition from shield stage at O'Higgins Guyot, Juan Fernández Ridge, Pacific SE. Deep Sea Res Part 1 Oceanogr Res Pap 141:33–42

Luhr, J, and Simkin, T (1993) Paricutin: A Volcano Born in a Mexican Cornfield: Phoenix, Geoscience Press, 427 p

Lipman PW, Sisson TW, Ui T, Naka J (2000) In search of ancestral Kilauea volcano. Geology 28(12):1079–1082

Masson DG, Le Bas TP, Grevemeyer I, Weinrebe W (2008) Flank collapse and large-scale landsliding in the Cape Verde Islands, off West Africa. Geochem Geophys Geosyst 9(7):Q07015

Maurya DM, Vasaikar S, Chamyal LS (2023) Geoheritage and geotourism sites from the Narmada-Son Fault (NSF) zone, Narmada River, and Mahi River in central mainland Gujarat, Western India. Geoheritage 15(4)

McCall GJH, LeMaitre RW, Malahoff A, Robinson GP, Stephenson PJ (1970) The geology and geophysics of the Ambrym Caldera, New Hebrides. Bull Volcanologique 34(3):681–696

McDougall I, Duncan RA (1980) Linear volcanic chains—recording plate motions? Tectonophysics 63(1–4):275–295

McDougall I, Embleton BJJ, Stone DB (1981) Origin and evolution of Lord Howe Island, Southwest Pacific Ocean. J Geol Soc Aust 28(1–2):155–176

McGuire WJ (2006) Lateral collapse and tsunamigenic potential of marine volcanoes. Geol Soc Spec Pub 269:121–140

Michael PJ, Zhang J, Trowbridge SR (2024) Earth's longest lava flows erupted from its largest igneous province: the submarine Ontong Java Plateau. Earth Planet Sci Lett 644:118905

Moore JG, Clague DA, Holcomb RT, Lipman PW, Normark WR, Torresan ME (1989) Prodigious submarine landslides on the Hawaiian ridge. J Geophys Res 94(B12):17465–17484

Mortimer N (2020) Zealandia (including New Zealand and New Caledonia). In: Encyclopedia of geology, vol 1–6, 2nd edn. Elsevier, Amsterdam, pp 631–641

Mortimer N, Parkinson D (1996) Hikurangi plateau: a cretaceous large igneous province in the Southwest Pacific Ocean. J Geophys Res Solid Earth 101(1):687–696

Mortimer N, Scott JM (2020) Volcanoes of Zealandia and the Southwest Pacific. N Z J Geol Geophys 63(4):371–377

Mortimer N, Campbell HJ, Tulloch AJ, King PR, Stagpoole VM, Wood RA, Rattenbury MS, Sutherland R, Adams CJ, Collot J, Seton M (2017) Zealandia: earth's hidden continent. GSA Today 27(3–4):27–35

Mossoux S, Saey M, Bartolini S, Poppe S, Canters F, Kervyn M (2016) Q-LAVHA: a flexible GIS plugin to simulate lava flows. Comput Geosci 97:98–109

Najita SY (2010) In the shade of the banyan tree. Contemp Pacific 22(2):349–361

Natland JH (1980) The progression of volcanism in the Samoan linear volcanic chain. Am J Sci 280-A:709–735

Natland JH, Turner DL (1985) Age progression and petrological development of Samoa shield volcanoes; evidence from K-Ar, lava composition, and mineral studies. In: Brocher TM (ed) Geological investigations of the northern Melanesian borderland, Circum-Pacific Council for energy and mineral resources earth science series. Springer, Berlin, pp 160–172

Neall VE, Trewick SA (2008) The age and origin of the Pacific islands: a geological overview. Philos Trans R Soc B Biol Sci 363(1508):3293–3308

Németh K, Cronin SJ (2009a) Phreatomagmatic volcanic hazards where rift-systems meet the sea, a study from Ambae Island, Vanuatu. J Volcanol Geotherm Res 180(2–4):246–258

Németh K, Cronin SJ (2009b) Volcanic structures and oral traditions of volcanism of Western Samoa (SW Pacific) and their implications for hazard education. J Volcanol Geotherm Res 186(3–4):223–237

Németh K, Gravis I (2022) Geoheritage and geodiversity elements of the SW Pacific: a conceptual framework. Int J Geoheritage Parks 10(4):523–545

Németh K, Kereszturi G (2015) Monogenetic volcanism: personal views and discussion. Int J Earth Sci 104(8):2131–2146

Németh K, Kósik S (2020) Review of explosive hydrovolcanism. Geosciences (Switzerland) 10(2):44

Németh K, Moufti MR (2023) Lava flow hazard and its implication in geopark development for the active Harrat Khaybar intracontinental monogenetic volcanic field, Saudi Arabia. Land 12(3):705

Németh K, Cronin SJ, Charley D, Harrison M, Garae E (2006) Exploding lakes in Vanuatu—"Surtseyan-style" eruptions witnessed on Ambae Island. Episodes 29(2):87–92

Németh K, Cronin SJ, Lolo F, Leavasa M, Solomona DS, Nelson F (2007) Volcanic evolution, oral traditions of volcanism of Western Samoa (SW Pacific) and their volcanic hazard implications. In: Geological Society of New Zealand and New Zealand Geophysical Society joint annual conference, Tauranga, NZ

Németh K, Cronin SJ, Smith IEM, Agustin Flores J (2012) Amplified hazard of small-volume monogenetic eruptions due to environmental controls, Orakei Basin, Auckland Volcanic Field, N Z. Bull Volcanol 74(9):2121–2137

Németh K, Fepuleai A, Muliaina T (2017) Samoa Geopark Project: a strategic plan to promote the volcanic geoheritage of Savai'i Island, Samoa. In: Conference: STAR 2017—the Pacific Islands science, technology and resources conference [26-29 June 2017] at: Nadi, Fiji. https://doi.org/10.13140/RG.2.2.19607.68009

Novković I, Dragićević S, Djurović M (2022) Geohazard and geoheritage. In: World regional geography book series. Springer, Berlin, pp 119–131

Nunn PD (1998) Sea-level changes over the past 1,000 years in the Pacific. J Coast Res 14(1):23–30

Okuma Y, Noda A, Koge H, Yamada Y, Yamaguchi A, Ashi J (2022) Surface friction of subducting seamounts influences deformation of the accretionary wedge. Tectonophysics 845:229644

Pacheco M, Cachão M (2021) Urban geology of Lisbon: the importance of the National Palace of Ajuda (Lisbon, Portugal). Geoheritage 13(3):84

Paris R, Bravo JJC, González MEM, Kelfoun K, Nauret F (2017) Explosive eruption, flank collapse and megatsunami at Tenerife ca. 170 ka. Nat Commun 8:15246

Pearce JA, Kempton PD, Gill JB (2007) Hf-Nd evidence for the origin and distribution of mantle domains in the SW Pacific. Earth Planet Sci Lett 260(1–2):98–114

Pekov IV, Zubkova NV, Agakhanov AA, Yapaskurt VO, Chukanov NV, Belakovskiy DI, Sidorov EG, Pushcharovsky DY (2017) Dravertite, CuMg(SO4)2, a new mineral species from the Tolbachik volcano, Kamchatka, Russia. Eur J Mineral 29(2): 323–330

Pérez-Torrado FJ, Paris R, Cabrera MC, Schneider JL, Wassmer P, Carracedo JC, Rodríguez-Santana Á, Santana F (2006) Tsunami deposits related to flank collapse in oceanic volcanoes: the

Agaete Valley evidence, Gran Canaria, Canary Islands. Mar Geol 227(1–2):135–149

Petchey FJ (2001) Radiocarbon determinations from the Mulifanua Lapita Site, Upolu, Western Samoa. Radiocarbon 43(1):63–68

Petterson MG (2020) Geo-tectonics, geology, and geo-resources of the Southwest Pacific. In: Encyclopedia of geology, vol 1–6, 2nd edn. Elsevier, Amsterdam, pp 724–741

Petterson MG, Neal CR, Mahoney JJ, Kroenke LW, Saunders AD, Babbs TL, Duncan RA, Tolia D, McGrail B (1997) Structure and deformation of north and central Malaita, Solomon Islands: tectonic implications for the Ontong Java Plateau Solomon arc collision, and for the fate of oceanic plateaus. Tectonophysics 283(1–4):1–33

Petterson MG, Babbs T, Neal CR, Mahoney JJ, Saunders AD, Duncan RA, Tolia D, Magu R, Qopoto C, Mahoa H, Natogga D (1999) Geological-tectonic framework of Solomon Islands, SW Pacific: crustal accretion and growth within an intra-oceanic setting. Tectonophysics 301(1–2):35–60

Price AA, Jackson MG, Blichert-Toft J, Hall PS, Sinton JM, Kurz MD, Blusztajn J (2014) Evidence for a broadly distributed Samoan-plume signature in the northern Lau and North Fiji Basins. Geochem Geophys Geosyst 15(4):986–1008

Price AA, Jackson MG, Blichert-Toft J, Konrad K, Bizimis M, Koppers AAP, Konter JG, Finlayson VA, Sinton JM (2022) Distinguishing volcanic contributions to the overlapping Samoan and Cook-Austral hotspot tracks. J Petrol 63(5):egac032

Prieto-Torrell C, Rodriguez-Gonzalez A, Aulinas M, Fernandez-Turiel JL, Cabrera MC, Criado C, Perez-Torrado FJ (2021) Modelling and simulation of a lava flow affecting a shore platform: a case study of Montaña de Aguarijo eruption, El Hierro (Canary Islands, Spain). J Maps 17(2):502–511

Ramalho RS, Quartau R, Trenhaile AS, Mitchell NC, Woodroffe CD, Ávila SP (2013) Coastal evolution on volcanic oceanic islands: a complex interplay between volcanism, erosion, sedimentation, sea-level change and biogenic production. Earth Sci Rev 127:140–170

Ramalho RS, Winckler G, Madeira J, Helffrich GR, Hipólito A, Quartau R, Adena K, Schaefer JM (2015) Hazard potential of volcanic flank collapses raised by new megatsunami evidence. Sci Adv 1(9):e1500456

Richmond BM (1992) Coastal geology of Upolu, Western Samoa. In: Keating BH, Bolton BR (eds) Geology and offshore mineral resources of the Central Pacific Basin. Springer, New York, pp 101–125

Richmond BM, Buckley M, Etienne S, Chagué-Goff C, Clark K, Goff J, Strotz L (2011) Deposits, flow characteristics, and landscape change resulting from the September 2009 South Pacific tsunami in the Samoan islands. Earth Sci Rev 107(1–2):38–51

Riou B, Chaumillon E, Chagué C, Schmidt S, Corrège T, Bujan S, Schneider JL (2024) Offshore evidence of historic and prehistoric tsunamis on the north shore of Tutuila (American Samoa). Sediment Geol 461:106572

Roche O, Phillips JC, Kelfoun K (2009) Pyroclastic density currents. In: Modeling volcanic processes: the physics and mathematics of volcanism. Cambridge University Press, Cambridge, pp 203–229

Rowley DB (2008) Extrapolating oceanic age distributions: lessons from the Pacific region. J Geol 116(6):587–598

Sand C, Baret D, Bolé J, Ouetcho AJ, Sahib M (2018) Sémoan settlement pattern and star mounds of Manono Island. J Polynesian Soc 127(1):91–109

Santos R, Quartau R, Brum da Silveira A, Ramalho R, Rodrigues A (2019) Gravitational, erosional, sedimentary and volcanic processes on the submarine environment of Selvagens Islands (Madeira archipelago, Portugal). Mar Geol 415:105945

Schmincke H-U (2004) Seamounts and volcanic islands. In: Schmincke H-U (ed) Volcanism. Springer, Berlin, pp 71–96

Schmincke HU, Sumita M (2016) Instability of oceanic volcanic edifices: examples of sector collapse, debris avalanches, and debris flows from Gran Canaria (Canary Islands). In: Submarine mass movements and their consequences, 6th international symposium, pp 605–616

Schneider-Christians D (1994) Schöpfungs-und Urzeitmythen aus Samoa [Creation myths from Samoa]. Anthropos 89(1–3):125–135

Seiuli BMS, Nikora LW, Awekotuku NT, Hodgetts D (2016) Fia Ola: grief recovery following a tsunami disaster in Samoa. J Pacific Rim Psychol 10:E5

Self, S, Sparks, RSJ, Booth, B, Walker, GPL (1974) The 1973 Heimaey strombolian scoria deposit, Iceland. Geological Magazine 111:539–548

Shreedharan S, Ikari M, Wood C, Saffer D, Wallace L, Marone C (2022) Frictional and lithological controls on shallow slow slip at the Northern Hikurangi Margin. Geochem Geophys Geosyst 23(2):e2021GC010107

Sinton JM, Sherrod DR (2020) Geology of the Hawaiian Islands. In: Encyclopedia of geology, vol 1–6, 2nd edn. Elsevier, Amsterdam, pp 742–757

Sischka L, Bosserelle C, Williams S, Ting JC, Paulik R, Whitworth M, Talia L, Viskovic P (2022) Reconstructing the 26 June 1917 Samoa tsunami disaster. Appl Sci 12(7):3389

Skowron C (1987) Western Samoa: a country profile prepared for "The Office of US Foreign Disaster Assistance Agency for International Development Department of State Washing, D.C," 20523, 42p

Smith WHF, Staudigel H, Watts AB, Pringle MS (1989) The Magellan seamounts: early cretaceous record of the South Pacific isotopic and thermal anomaly. J Geophys Res 94(B8):10501–10523

Song Q, Dong Y, Sun M, Hu X, Chen F, Ji Y, Wang J, Huang Z, Li H (2023) The types, grades and distribution features of geoheritage in China: an inventory. Int J Geoheritage Parks 11(3): 514–526

Spampinato L, La Spina A, Miraglia L, Morgavi D, Messina D, Maugeri SR, Giudice G, Calvari S (2024) Complex lava tube networks developed within the 1792–93 lava flow field on Mount Etna (Italy): insights for hazard assessment. Front Earth Sci 12:1448187

Stearns HT (1944a) Geology of the Samoan Islands. GSA Bull 55(11):1279–1332

Stearns HT (1944b) Geology of the Samoan Islands. Bull Geol Soc Am 55(11):1279–1331

Stearns HT (1946) The geology of the Hawaiian islands. Territory of Hawaii Division of Hydrography Bulletin 8:106

Tejada MLG, Mahoney JJ, Duncan RA, Hawkins MP (1996) Age and geochemistry of basement and alkalic rocks of Malaita and Santa Isabel, Solomon Islands, southern margin of Ontong Java plateau. J Petrol 37(2):361–394

Thompson GM, Malpas J, Smith IEM (1998) Volcanic geology of Rarotonga, southern Pacific Ocean. N Z J Geol Geophys 41(1):95–104

Thordarson T, Larsen G (2007) Volcanism in Iceland in historical time: volcano types, eruption styles and eruptive history. J Geodyn 43(1):118–152

Timm C, Hoernle K, Werner R, Hauff F, den Bogaard PV, Michael P, Coffin MF, Koppers A (2011) Age and geochemistry of the oceanic Manihiki Plateau, SW Pacific: new evidence for a plume origin. Earth Planet Sci Lett 304(1–2):135–146

Tonini R, Armigliato A, Tinti S (2011) The 29 September 2009 Samoa Islands tsunami: simulations based on the first focal mechanism solutions and implications on tsunami early warning strategies. Pure Appl Geophys 168(6–7):1113–1123

van de Lagemaat SHA, Kamp PJJ, Boschman LM, van Hinsbergen DJJ (2023) Reconciling the cretaceous breakup and demise of the Phoenix Plate with East Gondwana orogenesis in New Zealand. Earth Sci Rev 236:104276

van Wyk de Vries B, Byrne P, Delcamp A, Einarson P, Göğüş O, Guilbaud MN, Hagos M, Harangi S, Jerram D, Matenco L, Mossoux S, Nemeth K, Maghsoudi M, Petronis MS, Rapprich V, Rose WI,

Vye E (2018) A global framework for the earth: putting geological sciences in context. Glob Planet Chang 171:293–321

Velmurugan A (2008) Chapter 1—The nature and characters of Tropical Islands. In: Sivaperuman C, Velmurugan A, Singh AK, Jaisankar I (eds) Biodiversity and climate change adaptation in Tropical Islands. Academic, New York, pp 3–30

Vilches M, Ureta G, Grosse P, Németh K, Aguilera F, Aguilera M (2022) Effusion rate estimation based on solidified lava flows: implications for volcanic hazard assessment in the Negros de Aras monogenetic volcanic field, northern Chile. J Volcanol Geotherm Res 422:107454

Walker GPL (1986) Koolau Dike Complex, Oahu: intensity and origin of a sheeted-dike complex high in a Hawaiian volcanic edifice. Geology 14(4):310–313

Walker GPL (1987) The dike complex of Koolau Volcano, Oahu: internal structure of a Hawaiian rift zone. US Geol Surv Profess Pap 1350(2):961–993

Walker GPL (1990) Geology and volcanology of the Hawaiian islands. Pac Sci 44(4):315–347

Wang G (2015) Geoheritage features in Xi'an, China: Cuihua rock avalanche likely originating from an ancient earthquake. Geoheritage 7(3):285–297

Wang L, Tian M (2014) A discussion on the development model of earthquake relic geopark—a case study of the Qingchuan earthquake Relic Geopark in Sichuan Province, China. J Cult Heritage 15(5):459–469

Wang G, Ye W, Lv Y (2019) Loess geoheritage and geohazard protective measures at Luochuan Loess National Geopark in NW China. Geoheritage 11(3):1089–1100

Wang M, Wang M, Guo Z (2023) Effects of seamount subduction on structural deformation of Hikurangi accretionary wedge: insights from discrete-element modeling. Mar Geol Quat Geol 43(1): 82–93

White JDL, Valentine GA (2016) Magmatic versus phreatomagmatic fragmentation: absence of evidence is not evidence of absence. Geosphere 12(5):1478–1488

White JDL, Schipper CI, Kano K (2015) Chapter 31—Submarine explosive eruptions. In: Sigurdsson H (ed) The encyclopedia of volcanoes, 2nd edn. Academic, Amsterdam, pp 553–569

Williams SP, Davies TR, Barrows TT, Jackson MG, Hart SR, Cole JW (2014) Flank-collapse on Ta'u Island, Samoan archipelago: timing and hazard implications. In: Landslide science for a safer geoenvironment: volume 3: targeted landslides, pp 583–588

Williams S, Paulik R, Weaving R, Bosserelle C, Chan Ting J, Wall K, Simi T, Scheele F (2021) Multiscale quantification of tsunami hazard exposure in a Pacific Small Island developing state: the case of Samoa. GeoHazards 2(2):63–79

Williams S, Weaving R, Bosserelle C, Paulik R, Ting JC, Wall K, Simi T (2022) Impact of modelled tsunamigenesis on inundation and risk exposure: observations of the 2009 tsunami in Samoa. In: Australasian coasts and ports 2021—proceedings, pp 1007–1013

Wohletz KH (1983) Mechanisms of hydrovolcanic pyroclast formation: grain-size, scanning electron microscopy, and experimental studies. J Volcanol Geotherm Res 17(1–4):31–63

Wohletz K (2002) Water/magma interaction: some theory and experiments on peperite formation. J Volcanol Geotherm Res 114(1–2):19–35

Wu L, Taniguchi K, Tajima Y (2021) Impact of climate change on flood hazard at airports on pacific islands: a case study of Faleolo international airport, Samoa. J Disaster Res 16(3):351–362

Wyn Hughes G (1978) The relationship between volcanic Island genesis and the Indo-Australian pacific plate margins in the eastern outer islands, Solomon Islands, south-west pacific. J Phys Earth 26:S123–S138

Zakharovskyi V, Németh K (2021) Qualitative-quantitative assessment of geodiversity of Western Samoa (SW Pacific) to identify places of interest for further geoconservation, geoeducation, and geotourism development. Aust Geogr 1(3):362–380

Zakharovskyi V, Németh K (2022) Scale influence on qualitative—quantitative geodiversity assessments for the geosite recognition of Western Samoa. Aust Geogr 2(3):476–490

Zann LP (1994) The status of coral reefs in South Western Pacific Islands. Mar Pollut Bull 29(1–3):52–61

Zhang XG, Zhang XJ, Liu C, Wang H, Liu XL, Wang LL (2022) Discovery and significance of volcanic earthquake geoheritage of Laoheishan Scenic Area in Wudalianchi Global Geopark. Acta Geosci Sin 43(1):111–122

Zimanowski B, Büttner R, Dellino P, White JDL, Wohletz KH (2015) Chapter 26—Magma–water interaction and phreatomagmatic fragmentation. In: Sigurdsson H (ed) The encyclopedia of volcanoes, 2nd edn. Academic, Amsterdam, pp 473–484

3 Geodiversity of Samoa

3.1 Geodiversity Calculation Methods

3.1.1 Geodiversity and Reason for Its Calculation

Nature itself classified in two categories: living and non-living. The living part of nature includes the whole range of diversity of flora and fauna together with the environment they occupy creates biodiversity (DeLong 1996; Sarkar 2002). In more clear terminology: "biodiversity/biological diversity is species, genetic, and ecosystem diversity in an area, sometimes including associated abiotic components such as landscape features, drainage systems, and climate" outlined by Swingland (2013). Definition of biodiversity has a lot of different variations, but in general it is covering population of different species in chosen environment expressed as a number density of specific species-specific features. On the other hand, non-living nature is covering all other properties of the environment. These properties together can be explained through the term of geodiversity, but the exact definition of geodiversity also broad and varies greatly from pure number density estimates of specific geological features (e.g., number of geological units within a specifically define hierarchy system) to systematics that also provide some valorization of the relative abundance of the abiotic elements (Zakharovskyi et al. 2022). The term of geodiversity has been refined for the last 40 years in more relevant to spatial variations of specific abiotic features, but on current state it is understood as an analogy to biodiversity (Brilha et al. 2018; Burnelli et al. 2023; dos Santos et al. 2020). It cannot be strict analogy to the terminology of biodiversity, as sometimes they can share some of the elements depend on descriptive information, for example the soil element. The element of soil cover is composition of geological and biological matter (mineral and organic parts), which making it act as transition zone between abiotic and biotic nature. Hence, the clear definition about geodiversity does not exist yet, like the clear (or at least clearer or more broadly accepted way) definition of biodiversity. Therefore, researchers most of the time outline the definition, which they, or their methods more in match.

Our definition of geodiversity is a description of the variations of the abiotic (non-living) nature. Then, this description is divided on two categories in respect of their significance within the abiotic nature such as major and additional. The major category includes geological and geomorphological elements, which are the core part of the abiotic environment, where geology describes its (environment) "composition" (Fig. 3.1) and geomorphology define its "shape" (Fig. 3.2). Together, major elements create different variety of the land types with different scales and properties from plains and hills to valleys and mountains

Second category of geodiversity is additional, which contain other abiotic elements and making influences on major elements (Fig. 3.3). It includes hydrology, climate, endogenic (tectonics) and exogenic (solar radiation) processes, biological and anthropological influences and soils. These elements do not create stable material unlike geological and geomorphological elements, but rather transforming surface forms and compositions. For example, river (hydrological element) weathers rock formation from the cliffs side and accumulate its particles in other locations. Wind and precipitation eroding hills provoking slides and collapses of hills. Endogenic processes are more complicated as it highly connected with geological and geomorphological elements, but once more they provoke volcanic eruptions and earthquakes, which changing the form of the surface bringing new material or shuffling existed. Exogenic element is more connected with influence of space energies on the surface of the Earth such as meteorite craters and solar radiation. Anthropogenic element includes everything made by people, which can be mining areas, cities especially with stone buildings and some cultural value reflected in the rock formation like Holy Mountains (Olympic or Sinai).

K. Németh, V. Zakharovskyi, *Geodiversity of Samoa*, Geoheritage, Geoparks and Geotourism,
https://doi.org/10.1007/978-3-031-87220-4_3

Fig. 3.1 The concept of "composition" or geological parameters within the geodiversity estimation. Image is from the Saleaula Lava Field in Savai'i (13°27′41.17″S; 172°19′6.88″W)

Biological element or rather footprint is similar to anthropogenic, it could be holes for birds nesting carved in the soft rock formation, or fossils as reflection of biotic life. Meanwhile, the most challenging elements are soils as they can be counted as transition zone between abiotic and biotic part of nature (Bétard and Peulvast 2019; de Almeida Rangel et al. 2019; Fossey et al. 2020; Lausch et al. 2019; Reverte et al. 2020; Santos et al. 2020; Zaady et al. 2021). Soils is natural formations, which is result of rock transformation due to metabolic processes of micro-biome (Nazarenko et al. 2006). Hence, geodiversity describes abiotic part of nature and contain two major elements (geology and geomorphology) and unlimited number of additional elements (hydrology, climate, soils...), which highly depend on the research requirements.

Development of the terminology of geodiversity, was an important task to outline the non-living nature and point out on the importance of its preservation and admiration (Brilha 2018; Brilha et al. 2018). Hence, the geodiversity highly connected with range of other terms and disciplines behind them, where the most known and probably most relevant to our society are geotourism, geoeducation, and geoconservation.

Geotourism defines geological areas, objects, sites as the important points of tourist attraction for esthetical, recreational, and educational purposes. Then, it already touches the term geoeducation, which has a clear aim to increase the awareness of the population about the elements and meaning of abiotic nature and its role in the total environment. It specifically, established to popularize geological and geomorphological knowledge for students in various educational levels, tourists including local and external to a region and the local population to establish protection and conversation. Geoconservation describes the object and/or areas of geological and geomorphological category, which is needed to be protected due to their uniqueness or importance for keeping environmental balance. However, these three terminologies can be applied to different geological and geomorphological environments that require very different approaches how to measure

Fig. 3.2 The concept of "shape" or morphological parameters within the geodiversity estimation. The image is from Savai'i Island's northern coastal road looking roughly toward south (from 13°27′48.36″S; 172°25′6.51″W). Note the gentle slope of the lava shield with distinct scoria cones in its dorsal setting

them—For example, places with active volcanos can be considered as important geoeducational and geoconservational aspect to provide vital information about the volcanic hazards of such geological processes can pose to the environment. While volcanoes for instance are strong candidates to utilize their geoheritage for natural hazard resilience programs, the activity of them often pose direct hazards for visitors hence their geotourism perspective is in a special place within the complex nature of niche or adventure tourism. The high attractiveness of volcanic processes despite their obvious hazards can attract visitations and putting them into a complex geoeducation and geotourism perspective commonly referred to as geological features "playing with fire" (Erfurt-Cooper 2011). Hence, geodiversity opens new opportunities to improve spheres of education, tourism, and environmental protection. To recognize important geological/geomorphological locations researchers utilized different methodologies and techniques, which we will discuss in the next section.

3.1.2 Geodiversity Methods Qualitative, Quantitative, Qualitative-Quantitative

Recognition of geologically and/or geomorphologically important locations is an important part of geodiversity estimation, which contain different methodologies depend on the goals and terminology recognized by researchers. On current era of technological development, especially in computer science, cartography, and remote sensing, lead the assessment of geodiversity into the sphere of Geographical Information Systems (GIS) and computer modelling and statistics. GIS allows researchers to calculate high amount of data to recognize geological/geomorphological locations of interest without direct observation giving way to apply some sort of "remote sensing" approach to express the geological diversity of a region. The expressed diversity then can form the basis of territorial planning including the identification of key locations can be the regions where significant geosites are located or areas where conservation efforts need to be

Fig. 3.3 Additional elements as potential parameters to count in geodiversity estimations such as a small scoria cone (in the middle view at location of 13°58′15.50″S; 171°34′49.91″NW. Note the rugged older volcanics in the back and the circular scoria cone in the mid-view

increased. Hence, nearly all methods for geodiversity assessment have been done in these days within GIS environment and presented in different type of models.

In general, geodiversity modelling outlined with three major methods: qualitative, quantitative, and qualitative-quantitative (Zwoliński et al. 2018). Qualitative methodology is based on the expert view and can be rejected by other researchers based on subjective understanding of specific elements. In this method the expert him/her-self deciding, what locations can be considered as the most important for education, tourism, and/or conservation hence it is not required to perform geostatistics through GIS calculations. Then two other models mostly presented in GIS environment and requires some expertise in spatial statistics and specific GIS software skills.

Quantitative method is purely based on counting of units for different elements included into assessment. The accuracy of this methodology strictly depends on the amount of the data accessible for researcher. One of the latest research projects was made by Barančoková et al. (2023), where they applied quantitative methodology on the Slovenské rudohorie Mountains in Slovakia. Researchers utilized grid scale with 5×5 km^2, and then calculated the number of units from different maps fallen into each cell. The data contain six partial indices with geological diversity, tectonic diversity, geomorphological diversity, soil diversity, hydrological diversity, and mineral resources diversity. However, they present one of the standard methodologies (Dias et al. 2021), while other researchers create different modifications for example not utilizing grid systems (Forte et al. 2018) or have different level of complications (Arifin et al. 2023). The main drawback of quantitative assessment is incomparability because of differences in amount and accuracy of accessible data.

Then qualitative-quantitative methodology combines to methodologies together, where expert creates evaluation system for each unit included into assessment and can be utilized for regions with limited amount of data. Methodologies are not separated one from another, but sometimes can be utilized for clarification and/or justification of the results.

Hence, the geodiversity assessment is an important part required for recognition of locations suitable for geological/geomorphological education, tourism, and/or conservation. They can be assessed utilizing one for three methods or their collaboration defined as qualitative, quantitative, qualitative–quantitative.

3.1.3 QQG Methodology: Its General Understanding

For assessment of geodiversity, we developed qualitative-quantitative assessment of geodiversity (QQG), which is highly connected to our definition of geodiversity and applied here to Samoa (Zakharovskyi and Németh 2021, 2022a, b; Zakharovskyi et al. 2022). The definition divides geodiversity on major and additional categories, with geological and geomorphological elements as the formal and other abiotic elements as additional values. Here, our methodology contains 8-point evaluation system, which has been applied for major geodiversity values (geology and geomorphology), where geological and geomorphological elements are multiplied one on another creating the specific value, which we named "global". This name has been chosen to emphasize its applicability for the whole Earth, as geological and geomorphological elements are parameters, which describes the main framework of the abiotic nature of the Earth and can be recognized in any place on Earth. Both, elements cannot be presented in their full extends, so we once more paying attention on their global accessibility, hence for geological evaluation system we utilized rock types of rareness based on research of Blatt and Jones (1975) (Table 3.1). This method based on a global estimate of the relative abundance of certain basic rock types for simplicity. The system valorizes rare rock types with higher score, hence their spatial abundance is modified or weighted with their relative abundance based on their global abundance values. In simple expression, rarer rock types score higher in their value table.

Geomorphology can be present with a range of different models derived from digital elevation models (DEM) with a great range of resolutions. From digital elevation data various parameters and their thematic maps can be generated including slope angle, raggedness, roughness, and others, but the main importance it is their categorization appliance for the whole world. Hence, we utilized DEM-based values and derivative elements to describe the morphodiversity of the region. Among the morphodiversity elements the slope angle map derived from DEMs is the simpler informative enough data that can be obtained in freely available global data sets such as Shuttle Radar Topography Mission (SRTM) (accessed 12 October 2024, and DEM can be downloaded here https://portal.opentopography.org/raster?opentopoID=OTSRTM.082015.4326.1), Advanced Land Observing Satellite (ALOS) PALSAR (accessed 12 October 2024, and DEM can be downloaded here https://asf.alaska.edu/datasets/daac/alos-palsar-radiometric-terrain-correction/) or other local digital elevation services. The general concept is that higher slope angle values express higher morphodiversity hence value for geodiversity estimates based on the concepts that steeper regions normally exposing greater diversity of geological elements, normally having larger number outcrops (Fig. 3.4) and form a greater process dynamism that can provide additional element values such as faulting, folding or gravity mass flow processes and their deposits (Fig. 3.5).

As a combination of the geological and geomorphological value system (Table 3.2), global geodiversity can be expressed.

From the DEM various derivative values can be extracted that form more sophisticated morphological characterization can be achieved. In geodiversity calculation different type of models can be used, and they have range of differences between each other, and highly depends on the aim of research (Zakharovskyi and Németh 2022a, b).

Roughness calculates the differences between neighbor cells and features, which those pixels describe, which in result create the rate of surface irregularity. Roughness is highly depending on topography and can be influenced with surface features like trees, building, terrain, and relief (Abbas et al. 2020).

Table 3.1 Percentage of rock types exposed on Earth's surface as function of geological age. According to the table extrusive and intrusive rocks are the rarest type, hence they are having the highest value from geological perspective (Blatt and Jones 1975)

	Crystalline				Sedimentary	
Eras	Extrusive	Intrusive	Metamorphic and "Precambrian"	Total		No. of usable data points
Cenozoic	4	0	0	4	33	290
Mesozoic	2	1	1	4	18	177
Palaeozoic	1	1	<1	2	13	117
Precambrian	0	6	15	21	1	173
Age unknown	1	1	1	3	1	26
Total	8	9	17	34	66	783

Fig. 3.4 Steep slopes normally associated with greater potential of larger number of geological outcrops accessible partially due to the susceptibility for mass movement (at 13°55′4.55″S; 171°34′36.41″W)

Ruggedness is a quantitative measurement of the differences in terrain known also as "heterogeneity" (Habib 2021; Riley et al. 1999). The parameter was used for in-habitat modelling prediction for species type, and paleoglacier studies (Quesada-Román et al. 2020, 2021).

The Topographical Position Index is based on parameter of the calculated pixel compared to the mean values of the surrounding pixels. Oten used to determine the position of the studied slope, and as additional parameter for landforms classification (Meten et al. 2015).

Geomorphon is a new parameter, which extracts 10 different landforms from the surface. The landforms recognition is based on the position of the 8 neighboring cells compared to the studied one, which can be the same altitude as the studied cell, higher than the studied cell, or lower than the studied cell (Jasiewicz and Stepinski 2013; Stepinski 2011).

Geomorphons can therefore be used as a good proxy to express the morphodiversity and combine it into a geodiversity estimates as their spatial distribution can offer relatively easy to manage systematics within GIS environments (Fig. 3.6).

The various terrain analysis maps while offer numerous opportunities to extract values from the morphology, there are still now common understanding what is their meaning within the geodiversity calculations. In this respect the slope angle map is still the simplest and most used morphodiversity information source to be incorporated in a geodiversity calculation.

In general, the morphodiversity estimates heavily rely on the spatial resolution of the digital elevation sources. The most common and freely accessible DEMs such as SRTM normally operates in the 30-m resolution that is also a widely available data set free of charge with a global coverage. There are better resolution data sets such as the ALOS-PALSAR 12.5 m resolution digital data but that is not having global coverage and there are parts on Earth where large sections are missing. Individual countries, such as New Zealand operates with good resolution (8 m) DEMs or in key loca-

Fig. 3.5 Steeper slopes are also associated with the potential to delineate zones with greater morphodynamic regions and their deposits including rock falls, gravity mass flows, soil creeping, or debris fan formation along with potential fault plains or within a structurally determined fluvial network to be exposed (13°38′16.10″S; 172°15′13.29″W)

tions even LiDAR source DEMs with 1 m resolutions (accessed 12 October 2024 https://www.linz.govt.nz/). This means that for global comparison the 30 m resolution is a best we can use for global comparison, but this resolution certainly unsuitable for terrain characterization in a scale beyond of this and not optimal for geotrail design or more local interest. For this research, we utilized ALOS-PALSAR DEM (accessed 12 October 2024 https://search.asf.alaska.edu) due to lack of local models.

Meanwhile, additional values reflect additional category of geodiversity elements. These elements can be different,

Table 3.2 Global geodiversity element values

Values rate of 8-points evaluation system	Global (main) value of geodiversity	
	Geological element	Geomorphological element
	Rock type and age	Slope angle (°)
1—The lowest	Sedimentary cenozoic (alluvium sediments)	0–11.25
2—Low	Sedimentary mesozoic	11.25–22.50
3—Low to middle	Sedimentary paleozoic	22.50–33.75
4—Middle	Metamorphic precambrian	33.75–45
5—Middle to high	Intrusive precambrian	45–56.25
6—High	Extrusive cenozoic (vulcanic heritage)	56.25–67.50
7—The highest	Extrusive mesozoic	67.50–78.75
8—Rare	Sed. (Precambrian), Meta. and Intr. (Cenozoic, Mesozoic, Paleozoic), Extr. (Paleozoic, Precambrian)	78.75–90

and they cannot be applied to different part of the globe, for example hydrology can be calculated only for areas, water objects are present. Hence, for additional values we created local geodiversity, which describes more in details the uniqueness of the assessed region. These additional values can be different and sometimes increasing the details and their resolution of the global elements. For example, geological element presented for global scale can be described as middle value, but it can contain some unique volcanic feature, which can increase its uniqueness for the local perspective. This is a particularly important aspects on regions where for instance the main lithology is something common rock types within simple morphological assets while the presence of a specific geological form clearly stands out from its otherwise globally "average" values. Hence, global value of geodiversity is utilizing geological and geomorphological elements using parameters applicable for the whole world, while additional elements add details to the global geodiversity and emphasize the uniqueness of the chosen region which reflects its local values. This can be very important for local geoeducation, geotourism and geoconservation strategies to develop and be part of a local land use management platform.

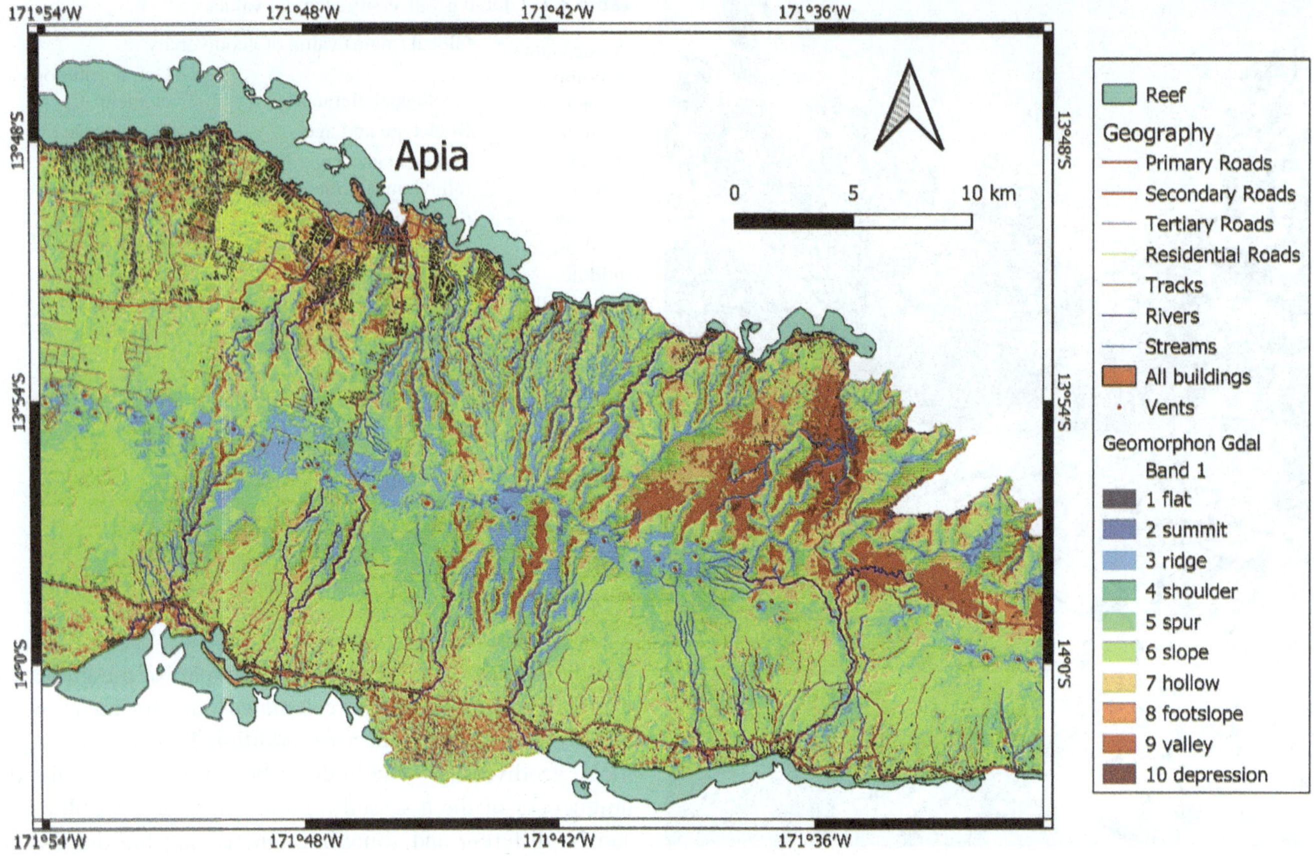

Fig. 3.6 Geomorphon map of central Upolu in Samoa showing the fine details of geomorphological features

3.2 Global Geodiversity of Samoa Islands

3.2.1 Introduction

The term of global geodiversity has been introduced in the previous section and tested in previous research, where we were thinking that some territories can be very homogeneous (the same value through the whole (most) area of research) from geological and/or geomorphological perspectives (Zakharovskyi and Németh 2022a, b). Especially, this issue is very apparent from the geological/geomorphological composition of Samoa Islands, which are formed with range of volcanic events of the same character (e.g., ocean island formation). It leads to the issue of recognition of locations with specific geodiversity, as commonly the whole territory suspected to contain the same value due to their similar geology and morphostructural assets. Alternatively, global geodiversity can also fall into high values if some geological or geomorphological features are rare and globally significant, regardless they may show limited variations in the small territory of islands. Hence, we decided to add more values, which can refine the result of assessment and highlight the location with higher uniqueness. Then, it led us to develop global and local values, where global value represents the rank of the studied region compared to the other regions throughout the world. It includes two parameters: geological element (geological architecture of the region expressed as the composition of the surface) and geomorphological element (form of the surface). Multiplication of these elements reflect the rank of the surface, which can be extracted from any point through the globe. However, in current assessment we are counting only surface composition excluding water objects as it is hard to study, due to the lack of information. Therefore, in subsections below we will explain in more details of global assessment of geological and geomorphological elements and their application for Samoa Islands with discussion of results.

3.2.2 Geological and Geomorphological Elements

The element of geology was the main point of geodiversity description, as this subject is coming as first during consideration and assessment of geodiversity as it is core parameter of the Earth's surface (Zakharovskyi et al. 2022). During the development of the geodiversity assessment, we concentrated on geology and its global application to find out some parameters of geology, which can be easy recognizable through-out the globe. Then, we understand that this should be based on rock types and age. These, bring us to the work of Blatt and Jones (1975) on "Proportions of exposed igneous, metamorphic, and sedimentary rocks". This research gives enough basic information to point out which type of rock, from which period is more rarely presented throughout the globe. This information can be used to establish a relative abundance ranking of various rock types and compare the study area, in this case Samoa, with the global pattern of rock type abundances. Then, all rock types have been ranked in an 8-point evaluation system (Table 3.3), where

Table 3.3 Local value of geodiversity ranking

Values rate of 8-points evaluation system	Local (additional) value of geodiversity	
	Volcanological element	Hydrological element
	Volcanic heritage and Eruptive centres	Strahler order
1—The lowest	Fagaloa volcanics	1st order
2—Low	Salani volcanics	
3—Low to middle	Vini tuff	2nd order
4—Middle	Mulifanua volcanics	
5—Middle to high	Lefaga volcanics	3rd order
6—High	Puapua volcanics	
7—The highest	Aopo volcanics	4th order
8—Rare	Additional 1 point for eruptive centres	

Sedimentary Cenozoic contain value 1, which is the lowest, sedimentary mesozoic—2 (low), sedimentary paleozoic—3 (low to middle), metamorphic precambrian—4 (middle), intrusive precambrian—5 (middle to high), extrusive cenozoic—6 (high), extrusive mesozoic—7 (the highest), the rest valued 8 points (rare): sedimentary precambrian, metamorphic and intrusive from different periods (cenozoic, mesozoic, paleozoic), extrusive paleozoic and precambrian. Then, 8-point ranking system has been applied to the geological model of Samoa Islands based on the available geological maps and geological reports (Zakharovskyi and Németh 2021, 2022a, b). Even, though geological map of Samoa Islands (Fig. 3.7) contains distinct lithostratigraphy (geological) units, nearly all of them fall under the same period (Pleistocene to Holocene) and rock type (various basaltic rocks or their volcaniclastic varieties), and one fall under Sedimentary Cenozoic (Alluvium Sediments). Hence, nearly the whole territory of Samoa Islands is represented as high value of geodiversity from geological element perspective (Fig. 3.8).

On the other hand, geomorphological element has different way of development. From, the beginning we wanted to consider geological types of landscapes, arranging them according to their uniqueness and give their values. However, we meet issues with its inapplicability throughout the globe, as there is no one specific system for creation of geomorphological maps, then different parameters and features can be named similar or some numerical parameters not equal

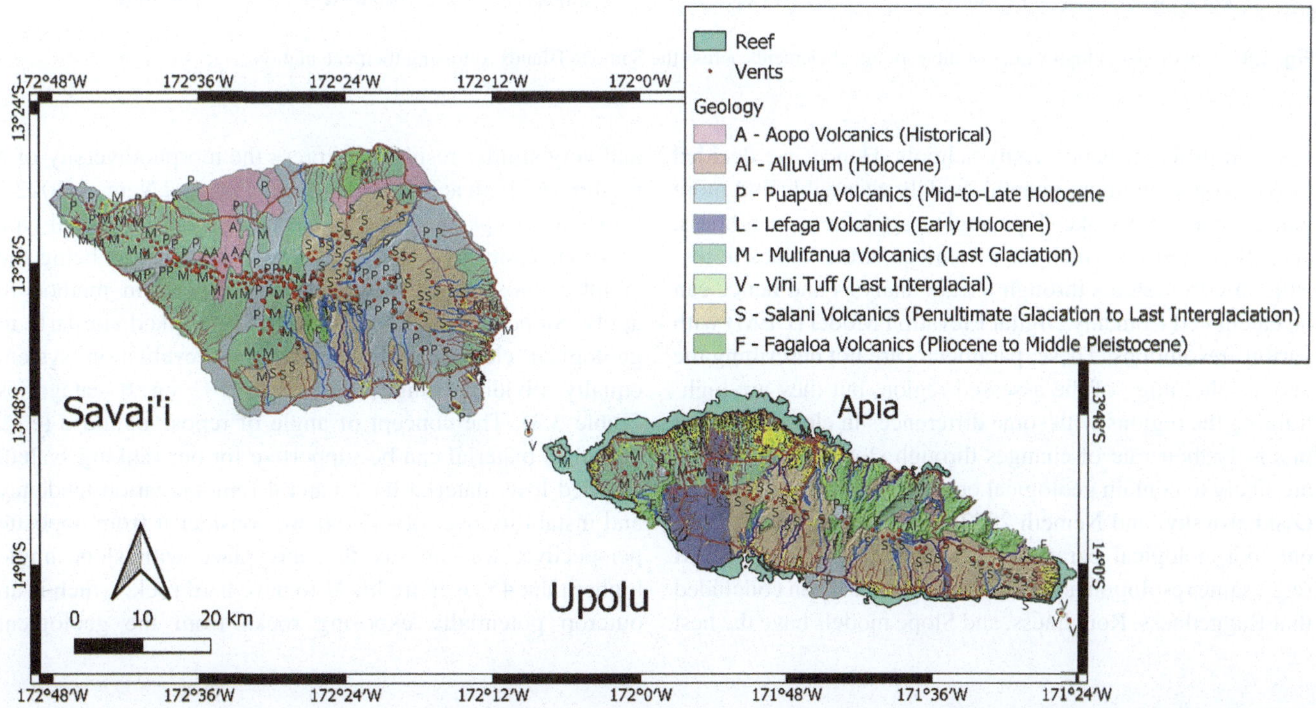

Fig. 3.7 Geological map of Samoa based on (Kear and Wood 1959; Keating 1991)

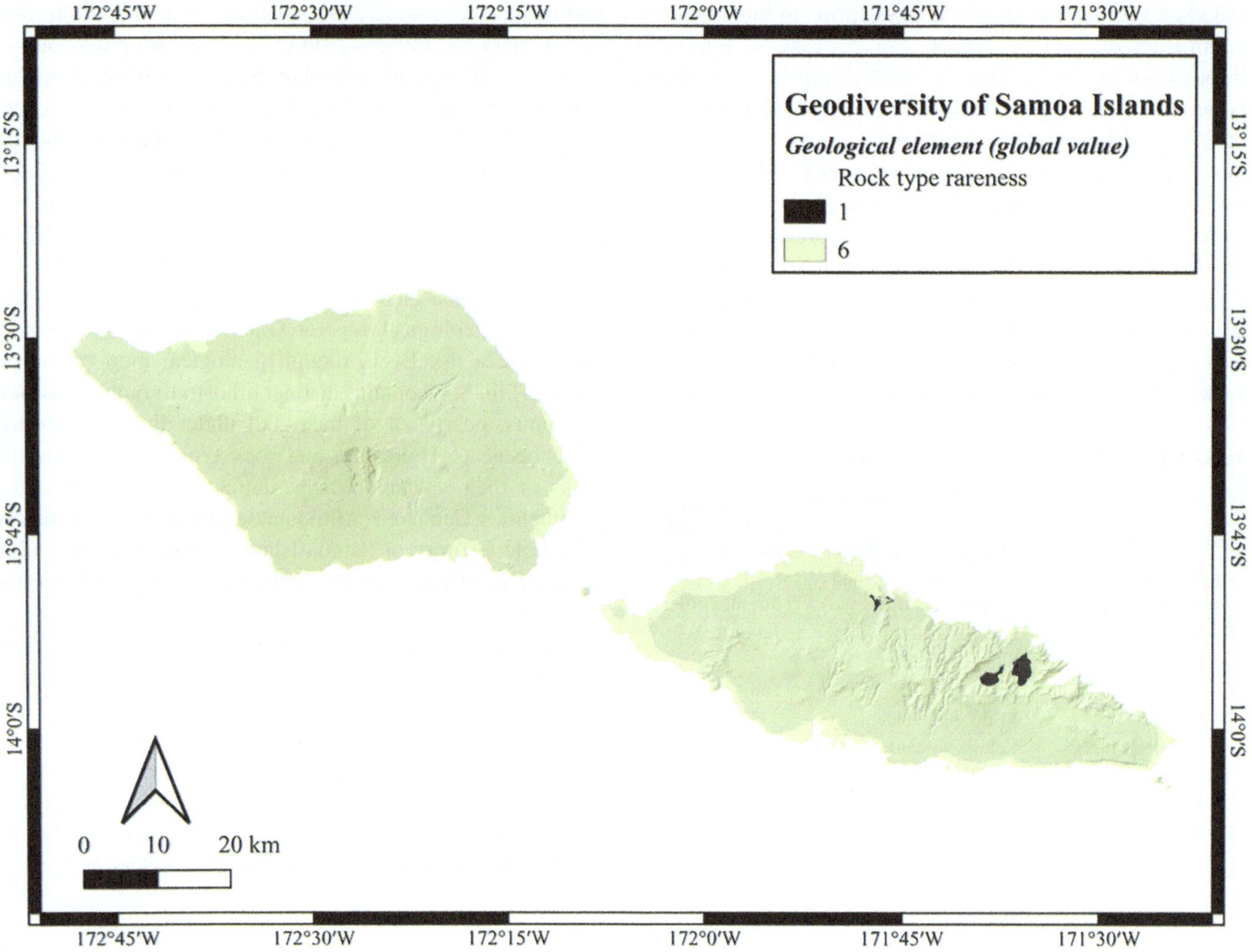

Fig. 3.8 Very even and high values of the geological elements across the Samoan Islands following their calculation

between different geodiversity schools. Hence, we decided to focus on some more general, widely accessible and more general parameters like slope angle, raggedness, roughness, and other models, which have some concrete values accessible in GIS systems through terrain analysis and hence can be calculated from any Digital Elevation Model (DEM) with various resolutions. These, parameters are not describing the geomorpholology of the assessed region, but they are highlighting the regions with some differences in elevation, most precisely their rate of changes through slope angles, which are likely to contain geological outcrop. In previous research (Zakharovskyi and Németh 2022a, b) on testing of 6 different morphological parameters applied for the same DEM (e.g., same resolution and data source), it has been concluded that Raggedness, Roughness, and Slope models have the best and very similar results to express the morphodiversity of a region with high accuracy (Zakharovskyi and Németh 2022a, b). Hence, slope angle model has be selected for global type of geodiversity assessment for Samoa Islands as being the simples, most robust hitherto easy to perform method to apply. Slope angle parameters has been ranked similarly to geological element within an 8-point evaluation system equally dividing range from 0 to 90° on 8 categories (Table 3.3). The concept of angle of repose of loose (e.g., granular) material can be supportive for our ranking system as piled lose material have natural remobilization tendency and instability over 40–45°. If we consider it from opposite perspective, we can say that any place with slope angle higher than 45° is more likely to have hard rock, which is an outcrop potentially exposing rocks from the geological

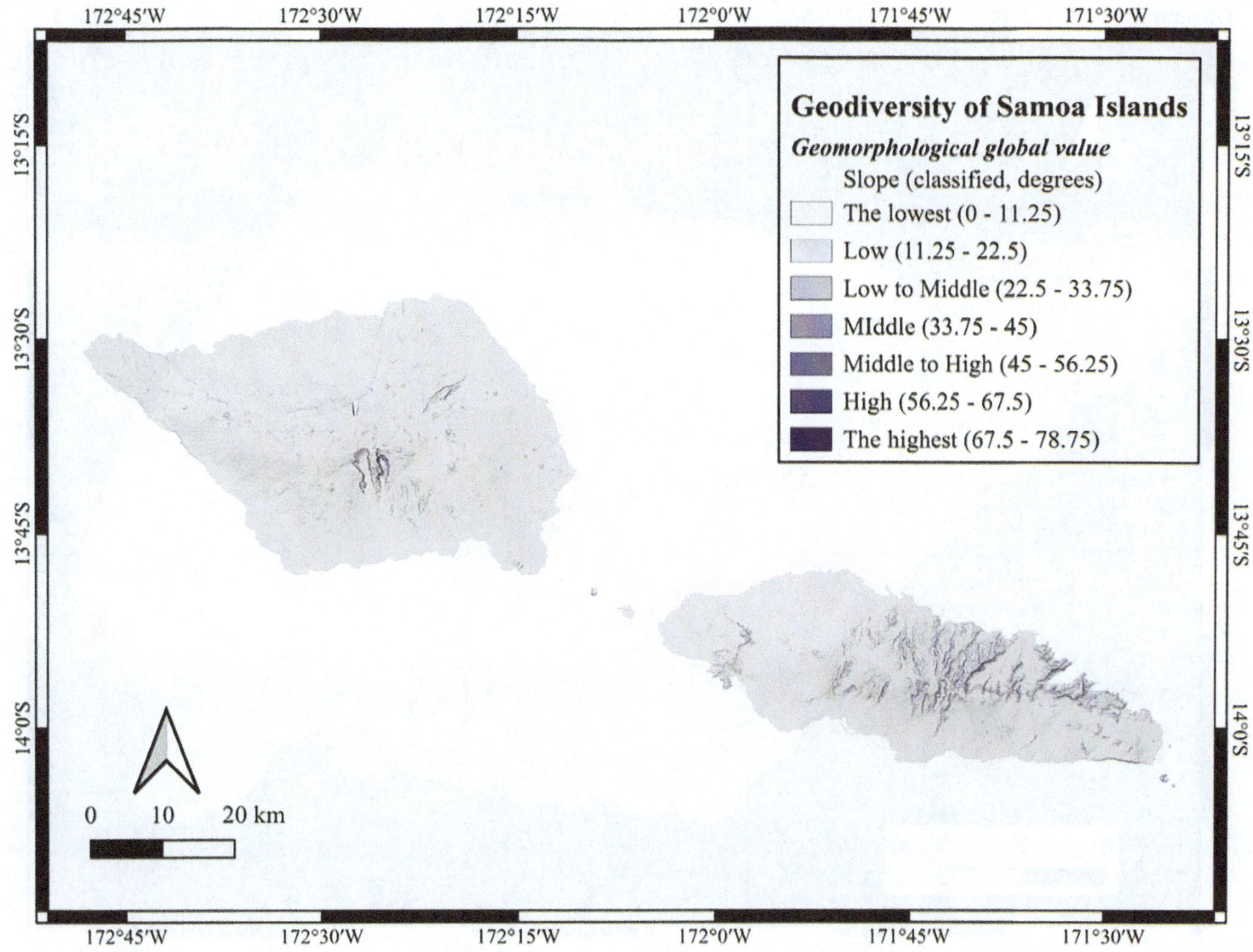

Fig. 3.9 Morphodiversity of Samoa

record. The slope angle categorization resulted a new map that expressing the morphodiversity of Samoa (Fig. 3.9). In this map the Samoa Islands general morphodiversity, or variations of morphology is relatively low beside the deeply incised valleys normally formed on the older surfaces (Figs. 3.9, 3.10, and 3.11).

Then, geological element has been applied for the Samoa Islands, which justify its homogeneity as both islands presented as basaltic shield volcanos.

Therefore, geomorphological and geological elements reflect global values for geodiversity and not reflecting the uniqueness of the assessed region in local level that is important for territorial planning. The global geodiversity in this perspective describes the geology and geomorphology but comparing the rock rareness and difference in forms with other locations throughout the globe. The result of assessment shows that geological element (rock rareness) by itself giving high score for nearly the whole territory of Samoa Islands (Fig. 3.8). However, it provides mostly homogeneous result, while inclusion of geomorphological element (slope angle) changing the picture and provides better allocation of the locations of interest (e.g., identification of geosites). They are mostly concentrated in the central-south part of Savaii Island and in the central-east part of Upolu Island (Fig. 3.9). In results, for Samoa Island the result of global assessment the recognition of the locations with higher values directly reflects the geomorphological element (Fig. 3.12) showing a uniform geodiversity pattern in a low-class value and some regions where morphology provides disruption on the monotonicity of the geodiversity estimates.

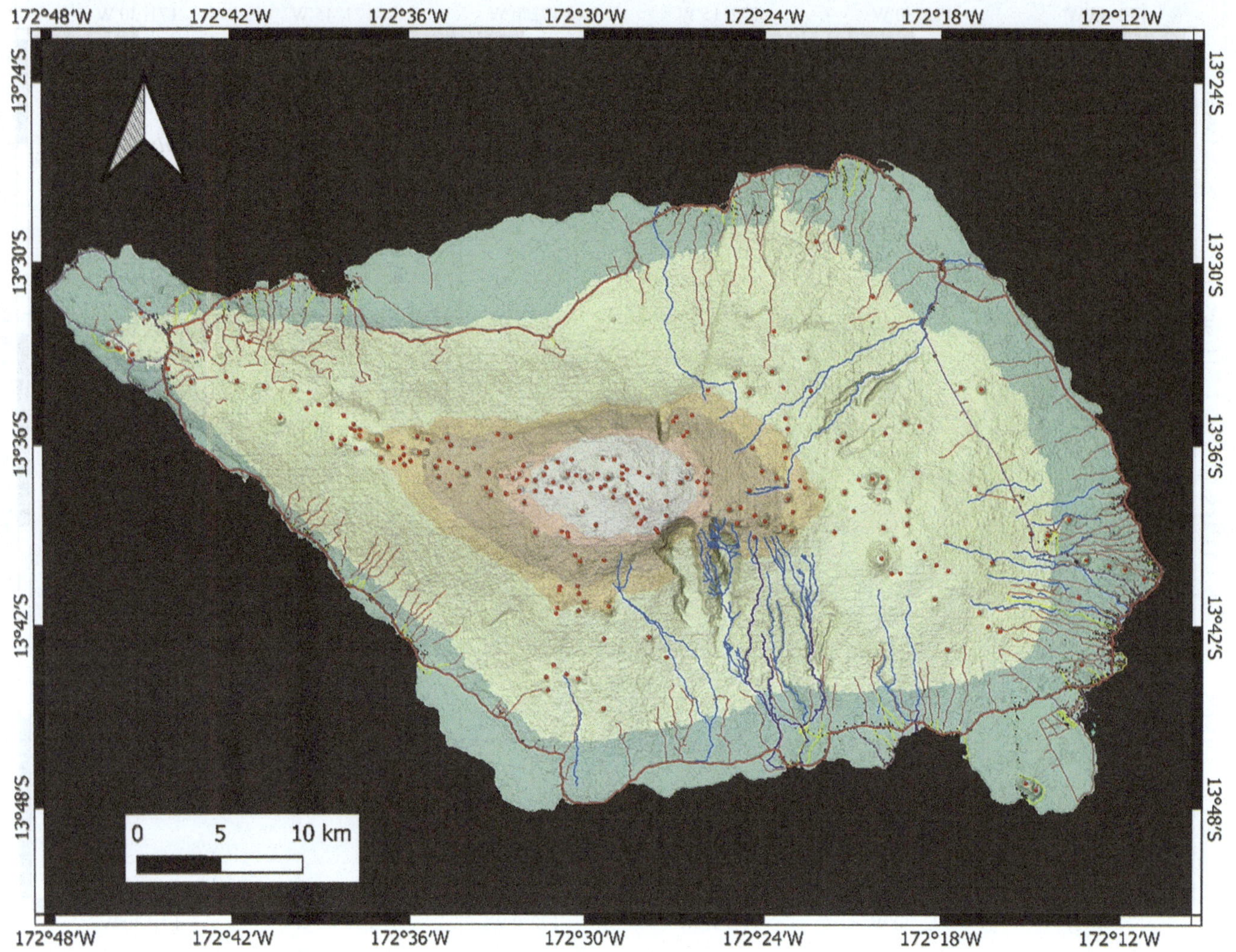

Fig. 3.10 Relatively low morphodiversity in the lava shield of Savaii on a relief map showing the changes of the slope of the landscape

Fig. 3.11 High morphodiversity in areas where deeply incised valley network is formed on the older surfaces (13°56′37.10″S; 171°36′12.52″W)

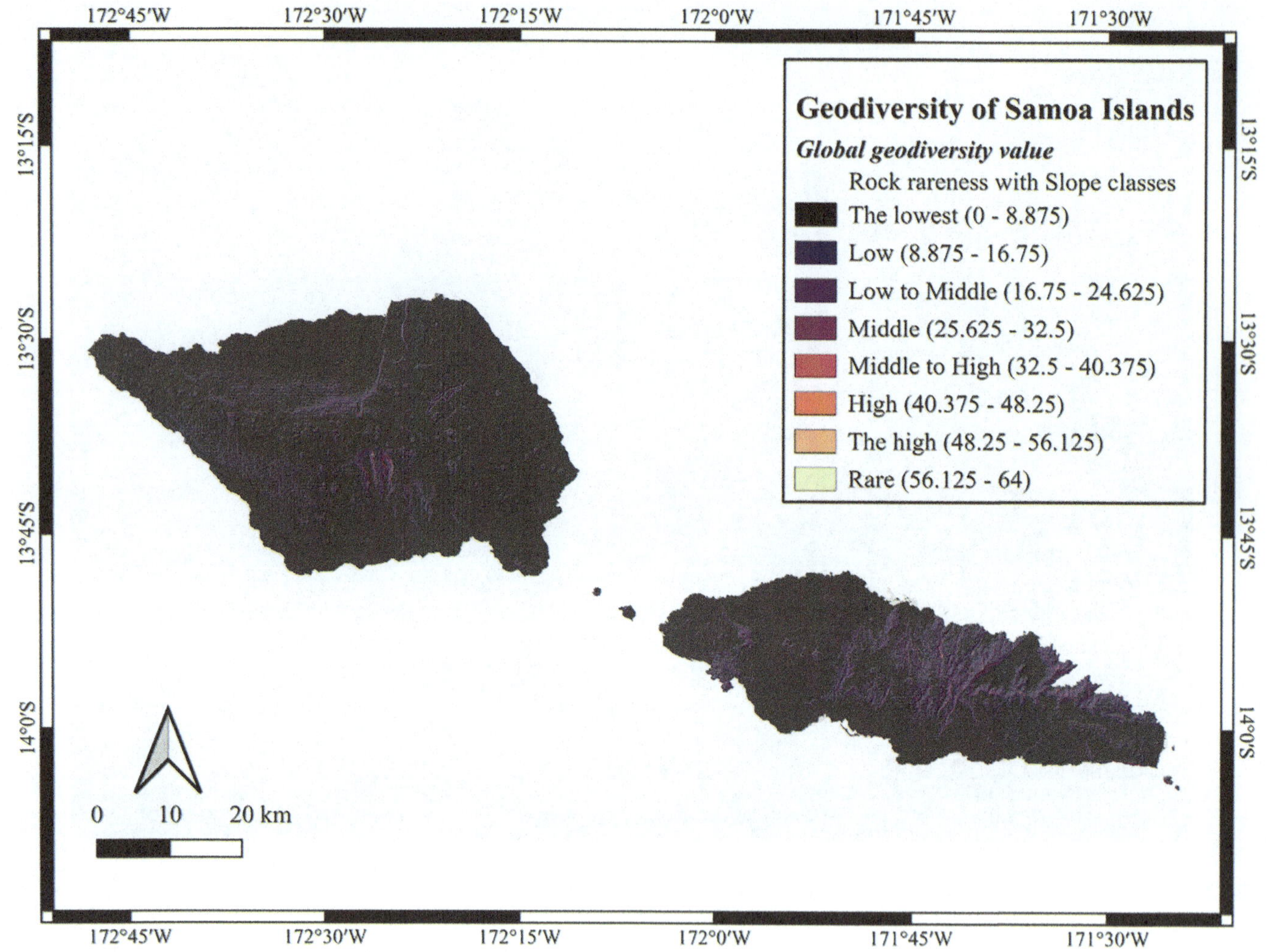

Fig. 3.12 Global geodiversity values of the Samoa Island

3.3 Local Geodiversity of Samoa Islands

3.3.1 General Concept

After, calculation of geodiversity based on global parameters, sometimes it shows very low amount of information to pursue further investigation to locate potential geodiverse regions (Table 3.3). Because of the limitation of the evaluation system and parameters chosen for assessment of geodiversity, the result is not bringing characteristically unique information but rather put the region under the global ranking system, which can be compared with other territories. This is a valuable result if we wish to see where Samoa Island stand in the global geodiversity perspective, but this way it is difficult to show the region geologically diverse nature locally as this information is vital for land use plannings. However, some regions can contain very low rank and/or highly homogeneous result, which bringing it to the point that the methodology purely oriented toward global perspective is not helping in highlighting the places of uniqueness inside the area of interest, in this case Samoa. To overcome this issue, we developed conception of local geodiversity, which contain all other information, which can be included into assessment to emphasize some specific uniqueness of the region.

First, these are additional elements of geodiversity such as hydrology, endogenic and exogenic geological processes, biological and anthropological footprints, climate and others. However, it is not presumably all other elements excluding geology and geomorphology, these two elements also can and should be improved with more additional information such as volcanic element, geomorphologically unique objects and so on. These additional elements are not chained to some specific rankings, and it can be different, based on the studied region. One specific parameter is important for assessment to keep it right is that additional elements must be evaluated according to 8-point evaluation system (like the

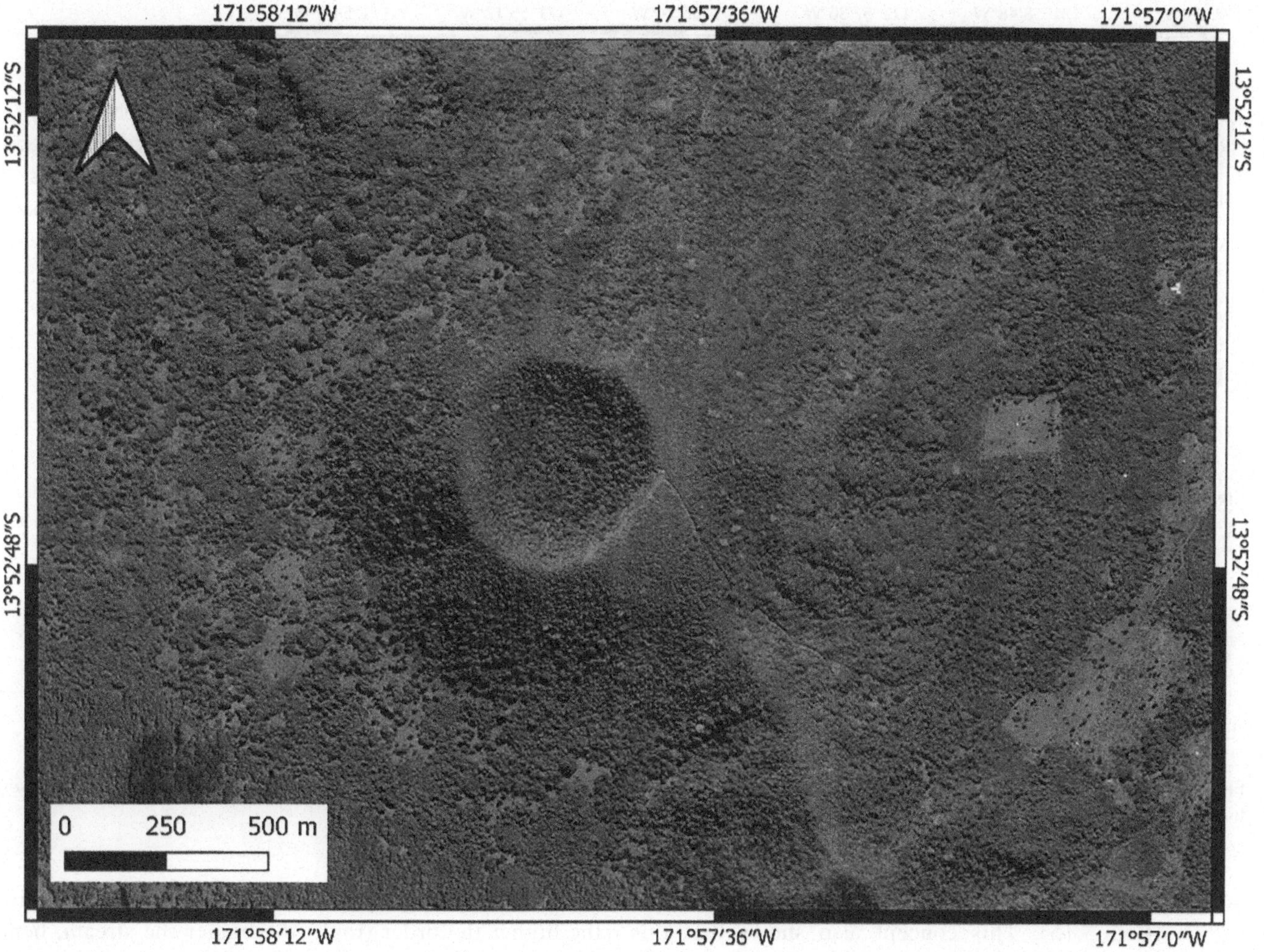

Fig. 3.13 The largest scoria cone (Tafua Upolu) of Upolu on a Bing satellite image. It is one of the youngest cone in Upolu and part of the Mid-to-Late Holocene Puapua Volcanics formed on the erosional remnant of a volcanic ridge of the Pliocene-to-Middle Pleistocene Fagaloa Volcanics (sharp ridge crossing diagonally the view)

global evaluation to keep the visualization simple and comparable) and then be added to global value of the studied region. Then, the more elements will be included into assessment the more accurate geodiversity model will be produced. However, when researchers add new elements to global values, which are unique for the studied region it automatically transforms it to local value as the result can be compared only for the places of the studied region. As geodiversity of Samoa Islands have homogeneity in their global value, volcano heritage (volcanic element) and hydrological element have been added to emphasize the result of assessment.

3.3.2 Volcanic Element

Volcanic geoheritage is an element of geodiversity (Fepuleai and Németh 2019), which has been first added into QQG assessment of geodiversity for Samoa Island to refocus the search of high value geodiversity region on the local scale. Volcanological element contain two parts, first is range of basaltic rocks ranked according to 8-point evaluation system and volcanic cones (dominated by scoria or cinder cones) recognized on satellite images (Fig. 3.13) and DEM (Fig. 3.14) of the region. The lowest value has been given to fagaloa volcanics—1 point, then salani volcanics—2 (low), vini tuff—3 (low to middle), mulifanua volcanics—4 (middle), lefaga volcanics—5 (middle to high), puapua volcanics—6 (high), aopo volcanics—7 (the highest). Meanwhile, the Rere statues have not been arranged to rock formation, but eruptive centres received 1 point for rock they formed on, and the centres located on aopo volcanis receive 8 points.

The volcanic elements in a local scale or in the local geological context immediately created a more geodiverse picture of the region indicating the importance of using an internally coherent and geologically validated

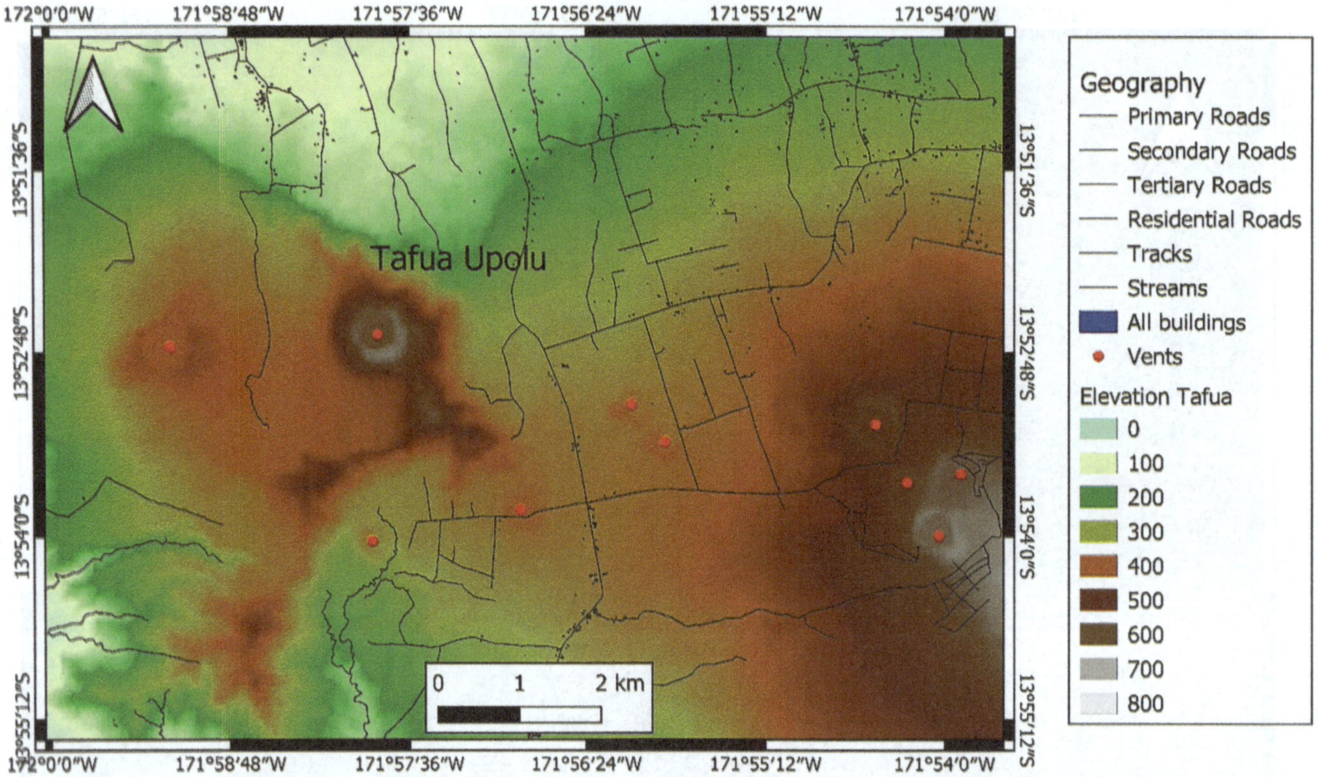

Fig. 3.14 Typical scoria cone field in western Upolu around Tafua Upolu Holocene cone on a DEM derived from ALOS-PALSAR 12.5 m resolution elevation data

conceptual framework to capture the geodiversity of a region (Fig. 3.15). This concept also shows that this approach can be applied more generally if the purpose of geodiversity measurements is driven by a need to compare to a region globally. In a simple expression, this approach allows to compare geodiversity within its own geological concepts, in the Samoan case within volcano geoheritage.

3.3.3 Hydrological Elements

Hydrological element is not as great for geodiversity assessment, but it is reasonably good for highlighting regions with valleys and some special features such as waterfalls (Fig. 3.16). To assess hydrology, Strahler order has been used as it can be easily extracted from DEM utilizing GIS software. Strahler order represents the rank of channels, where each starting point is the first channel, then places of merging of channels with the same ranks creates a new channel with next rank (Devroye and Kruszewski 1996). Channels of different ranks after merging not changing the rank. In a simple explanation, the higher the order (rank) the bigger the stream, hence higher rank assumed to carry more geological information through its sedimentary particles it transported from the catchment areas. Ranks than also applied to an 8-points evaluation system and added to the global value of geodiversity (Fig. 3.17). This addition increases the value of the region located along valleys of streams that is defined here as the representation of the local hydrological element. In addition, hydrological elements can also be emphasized with some other water objects such as: lakes, waterfalls and so on. Hence, Strahler order has been calculated utilizing DEM of Samoa Islands (Fig. 3.17). Interestingly this approach shows elevated geodiversity values in those areas located in older and commonly erodible terrains where valley incision, and surface run off water feed the major river network of the islands.

In conclusion, local geodiversity is more emphasizing the uniqueness of the studied region, where its accuracy highly depends on the amount of information included into assessment. For geodiversity assessment of Samoa Islands local

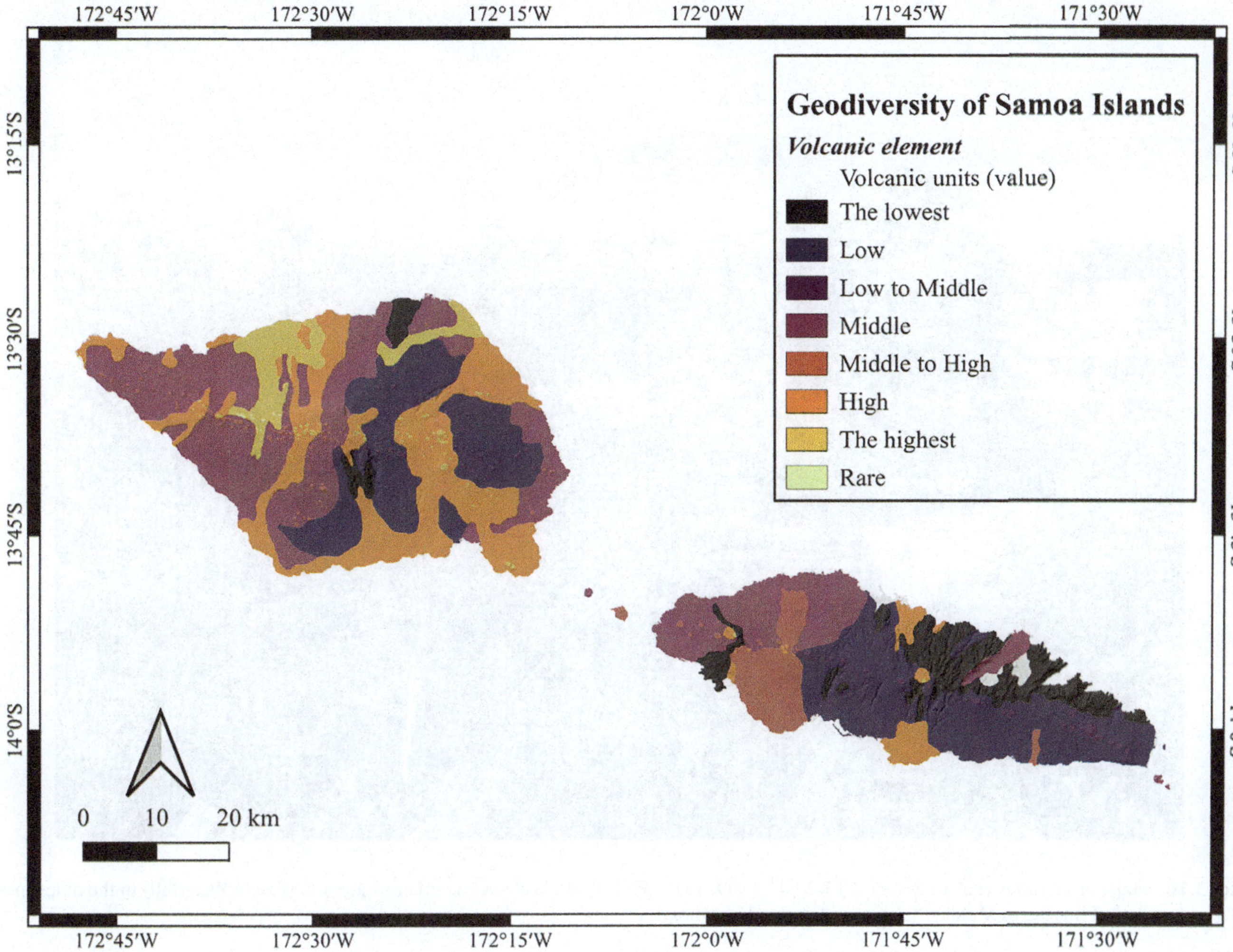

Fig. 3.15 Geodiversity estimates capturing the volcanic geodiversity of Samoa

geodiversity is a good solution to avoid the issue with homogeneity of the global values for this region.

3.3.4 Coastal Elements

Samoa main touristic attractions are centered along its spectacular coastal settings. The older island of Upolu is surrounded by extensive coral reefs creating environments of shallow marine conditions sometimes several kilometers wide (Figs. 3.18 and 3.19) while in areas where terrigenous input is significant by local rivers, mud flats, mangrove fringed regions formed (Fig. 3.20), making the region coastal diversity great. Coastal geodiversity (including the shallow marine geodiversity) is a subject that is in interest, but no universal approach developed yet what and how should be included in any calculations. If we consider diversity as a number density of variations of coastal environments some estimation of the variations is possible. However, to add value for this such as ranking coast types can be problematic. Steep cliffs in other hand that are visually spectacular (Fig. 3.21) but commonly difficult to access (hence not so useful for geotourism) can expose older stages of the island grows and some geomorphological features maybe indicative of recent sea level changes.

Samoa coastal diversity has been tested here to see how coastal diversity may influence the overall geodiversity estimates. In this calculation we defined three main coastal types such as (1) coastal cliffs, an area where steep rocky coast forming a narrow bend with extensive outcrops, (2) coral sand beaches formed behind fringing reefs and (3) mangrove dominated muddy plains formed normally behind

Fig. 3.16 Major river valley (Falefa River) at 13°53′43.37″D; 171°35′22.06″W) with characteristic water fall (Falefa Waterfall) in the oldest part of Upolu Island (Pliocene to Middle Pleistocene Fagaloa Volcanics)

complex lava flows reached coastal regions and associated with extensive river input carrying muddy sediments. Each of these zones are additional values with the highest given to coastal cliffs, medium value for the coral sand beaches while the lowest but still additional values provided for the mud flats with mangroves. The additional value to the coastal regions is justified based on their existence in regions normally not associated with specific highly ranked geological regions nor showing dramatic morphological variations. The identified coastal sections then assigned a specific buffer area to express the potential spatial extension where the coast need to be considered as an important additional factor for the geodiversity estimate. Here we applied a 500 m seaward and inland buffer for the mud flats and coral sand beaches, while 200 m for the cliffs. This buffer zones are calculated from the low tide line available for the Samoa Islands. The coastal geodiversity then shown on the map (Fig. 3.22).

3.3.5 Anthropogenic Elements

While anthropological elements are in question whether they should be included in geodiversity calculation or not, here we list few key points that make the Samoa Islands special in these respects and probably the anthropogenic elements may need to be consider. Here we understood anthropogenic elements are those that are connected to the human society either as tools, form or living or traditions including social activities that may be associated to the abiotic nature. This later can be an important element, as in Samoan culture even the deep cultural aspects of tattooing or dancing has clear link to the

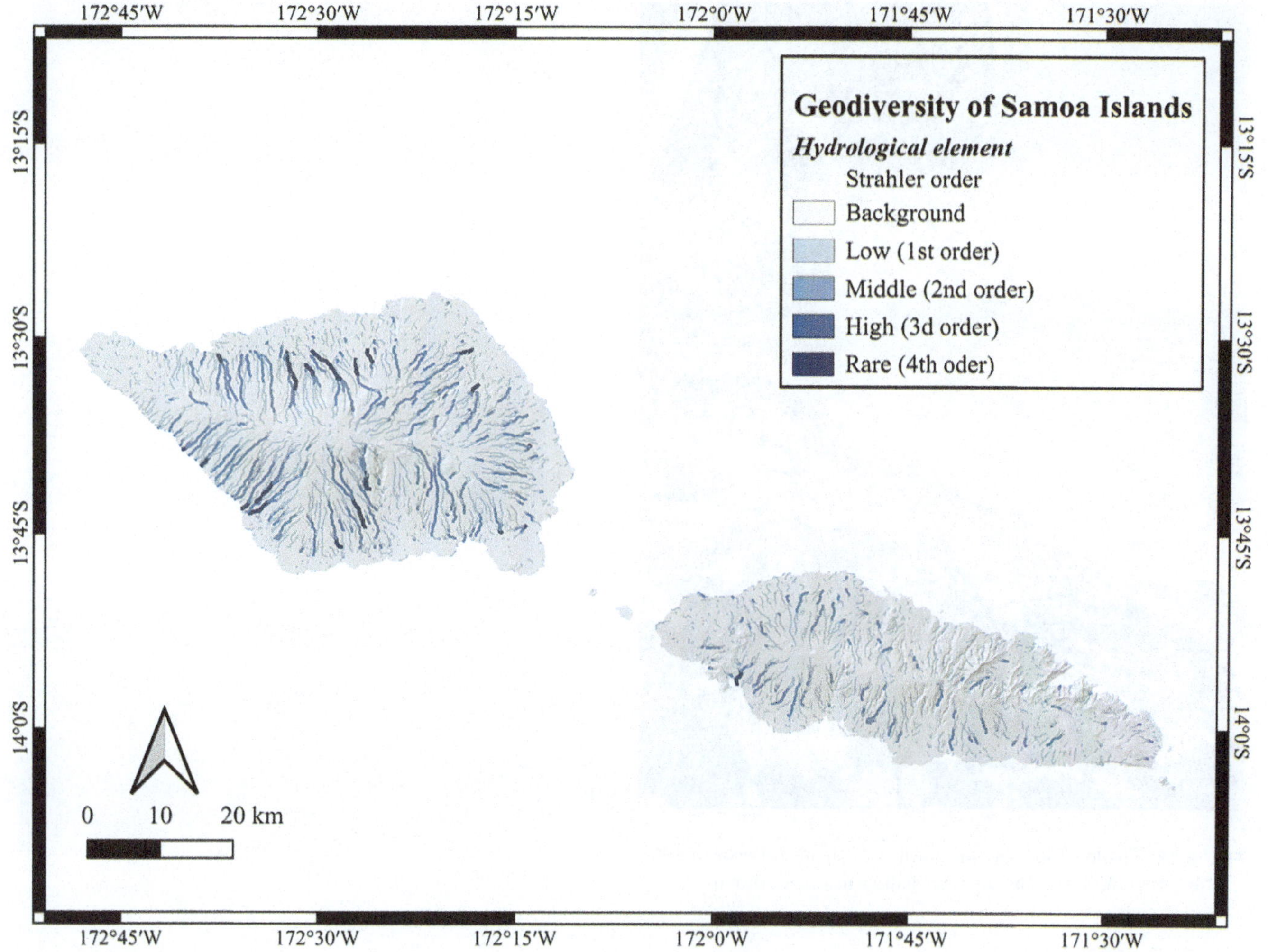

Fig. 3.17 Hydrological elements as part of the geodiversty estimation of Samoa applying Strahler order concept. Eventually, the higher the order the more diverse geological information associated with the hydrological element hence the addition to the overall geodiversity values are greater

abiotic nature they evolved. The problem to factor those elements into any geodiversity estimates is that they are not obviously linked to a specific location hence their spatial aspects is uncertain, or we still don't have tools how to collect that information and how to link them to specific places. Somehow easier to visualize the geodiversity aspects of various archaeological sites such as stone fences (Fig. 3.23), star mounds (*fetu ma'a*) (Fig. 3.24) or tombs (Fig. 3.25) or other artifacts associated with ancient or modern communities and linked to the island's geological assets (Fig. 3.26).

The problem of dealing with the anthropogenic assets in geodiversity estimates not only associated with the limited knowledge of their locations, but also how to incorporate their locations in a GIS environment and how we rank them. One way is to locate them and use some sort of buffer to express their spatial presence. Probably an average of 250 m radius buffer circle could express any of the point features special impact reasonably well by considering that within that area a specific feature has their own impact to the associated human community. It is more difficult to deal with linear features such as stone walls as there are so many, but nearly no available data exists on their special distribution. In this perspective we are not adding stone walls in the geodiversity calculations. Other important geoarchaeological sites as tombs, star mounds, human occupational sites all together treated as a point sources with high additional values and added to the global geodiversity values calculated earlier (Fig. 3.27).

Fig. 3.18 Shallow laguna in north of Upolu (13°50′14.59″S; 171°44′36.84″W) form a broad calm shallow marine environment

Fig. 3.19 Reef fringed beach in S Upolu with coral sand and lava flow lobes (Matareva Beach at 13°59′21.06″S; 171°55′57.13″W), as popular tourist destination

Fig. 3.20 Mangrove fringed coastal area with quite bays formed behind lava flows in Savai'i Island (13°28′12.94″S; 172°25′28.79″W), providing geodiverse coastal regions with associated elevated biodiversity

Fig. 3.21 Coastal cliff region in older tuff cones of Fanuatapu Island in eastern Upolu (14°0′47.77″S; 171°24′3.39″W)

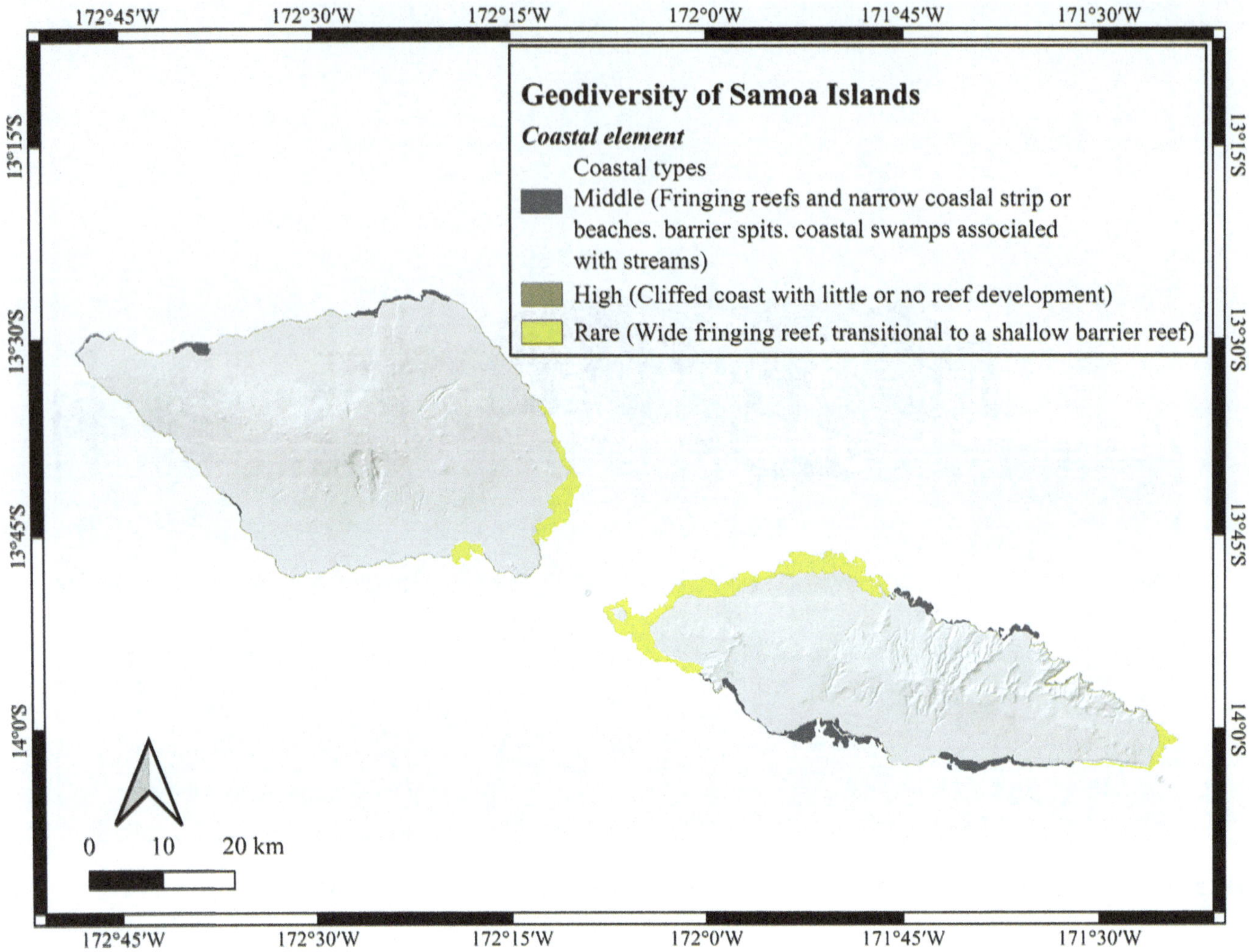

Fig. 3.22 Coastal diversity values calculated for the Samoa Islands

Fig. 3.23 Extensive stone walls made out basalt boulders collected from a village plantation in northern Savai"i Island

Fig. 3.24 Tombs made out basaltic fragments clearly show the geocultural aspects of the basalt as a potential heritage stone of Samoa (13°50′11.59″S; 171°47′10.12″W)

Fig. 3.25 One of the large star mounds of Samoa in the Palauli District near the village of Vaitoomuli, Savaii (**a**) and the Pulemelei Pyramid (**b**) also on Savai'i on a Google Satellite image

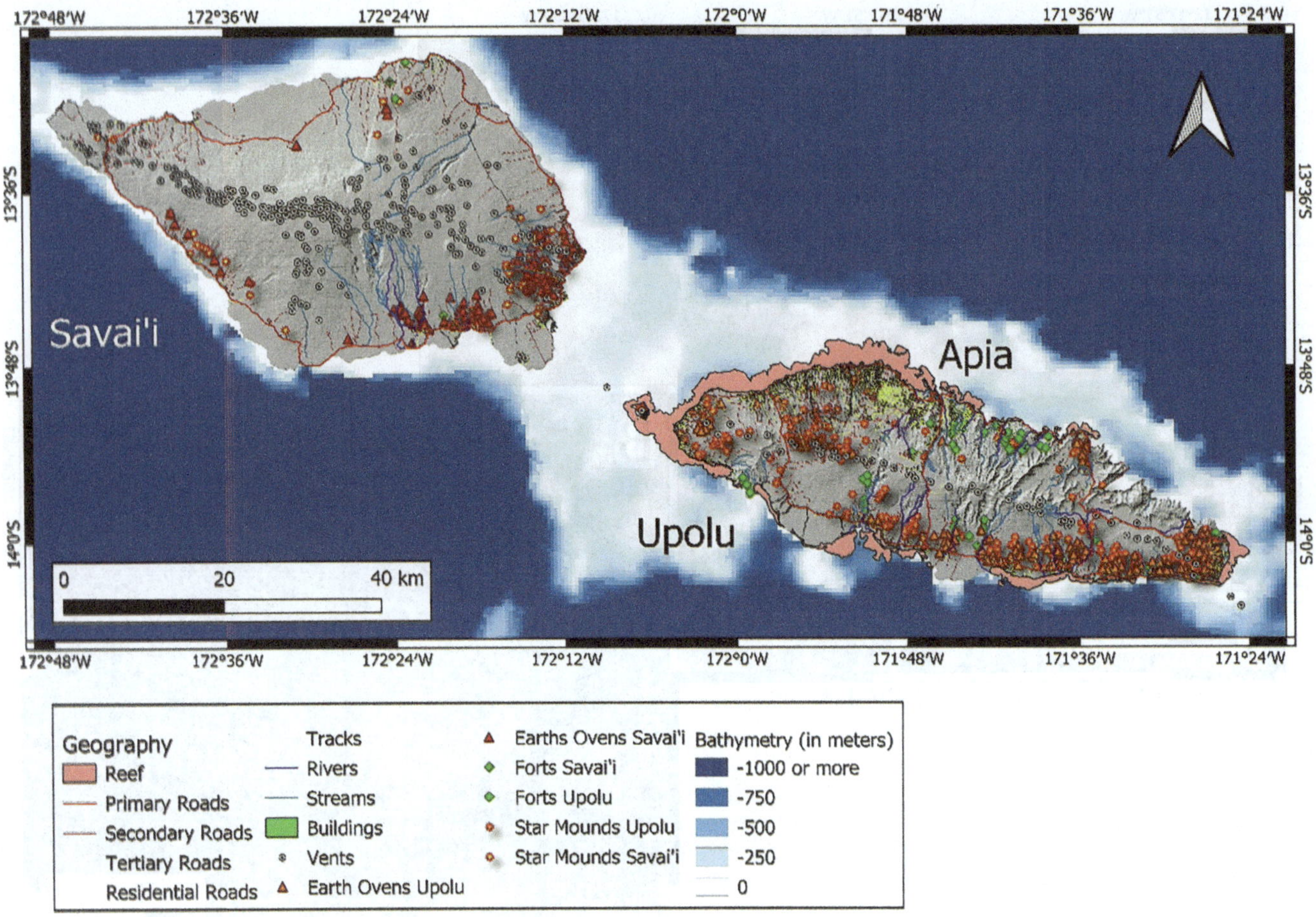

Fig. 3.26 Location of key anthropogenic elements that are in strong relationship with the geology and can be considered as part of geodiversity such as Earth Ovens, Forts and Star Mounds, all forming significant geocultural landmarks build from basaltic rocks. Data is based on the UTU: Samoa archaeology and cultural heritage database (Jackmond et al. 2022)

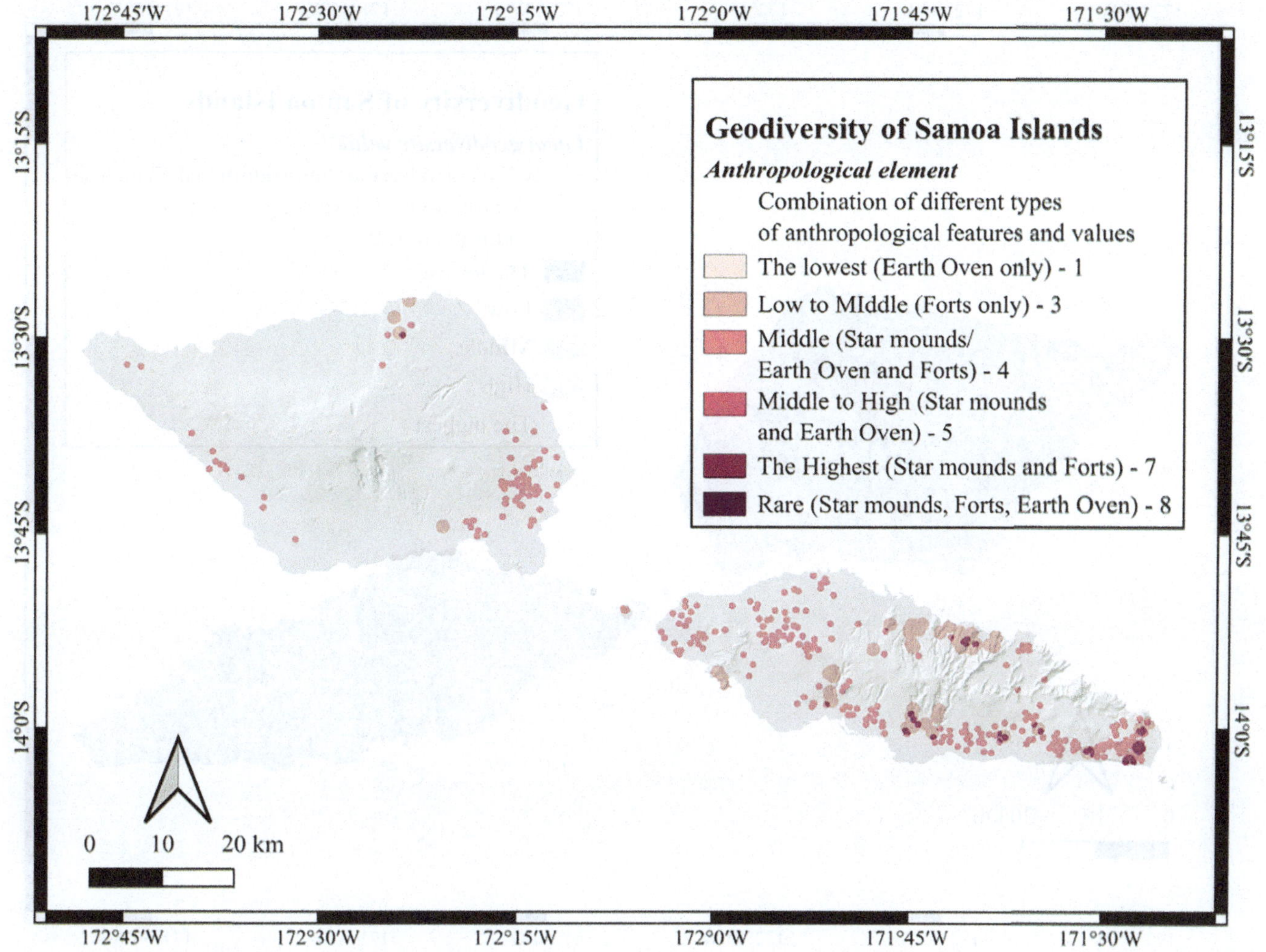

Fig. 3.27 Diversity estimates of the geocultural elements considered within the geodiversity estimation

3.4 Geodiversity Results and Their Discussion

Geodiversity of Samoa is complex and can be divided into global and local elements. Global elements are those that can be compared with other locations on Earth along their global geomorphological assets as well as the relative abundance of the exposed (mapped) rock types and geological formations. The geodiversity calculations using global elements provided a relatively modest diversity attribute. This is in good agreement with the islands shield-like morphological features dominated with gentle sloping hillsides in most of the regions, especially along its young island of Savai'i. In addition, the islands almost entirely composed of the most common volcanic rock types, basalt (in broad sense). The non-volcanic rock types are Quaternary beach and alluvial sediments. The older part of the island however has some complex morphology and through it exposed older rock successions associated with the early shield growth, erosion, and rejuvenation of ocean island volcanism.

Samoa local geodiversity is more complex and associated with deep geocultural aspects both through archaeological remains and active cultural traditions (Fig. 3.28). In the calculations presented here, the local geodiversity elements were enriched by the young volcanic edifice inclusion that is a dominating landscape element across both islands, however their small surface and volumetric features preventing them to add significant diversity value for the total geodiversity estimates (Fig. 3.28). The hydrology (through short and relatively simple drainage system, alluvial plains provided another addition elements to be included in local geodiversity estimates. The island rich geocultural features such as location of various archaeological sites were another key aspect to develop the total geodiversity estimate of the island. Overall, Samoa is a prime location to show the complexity and difficulty to provide geodiversity estimates in areas like islands where the

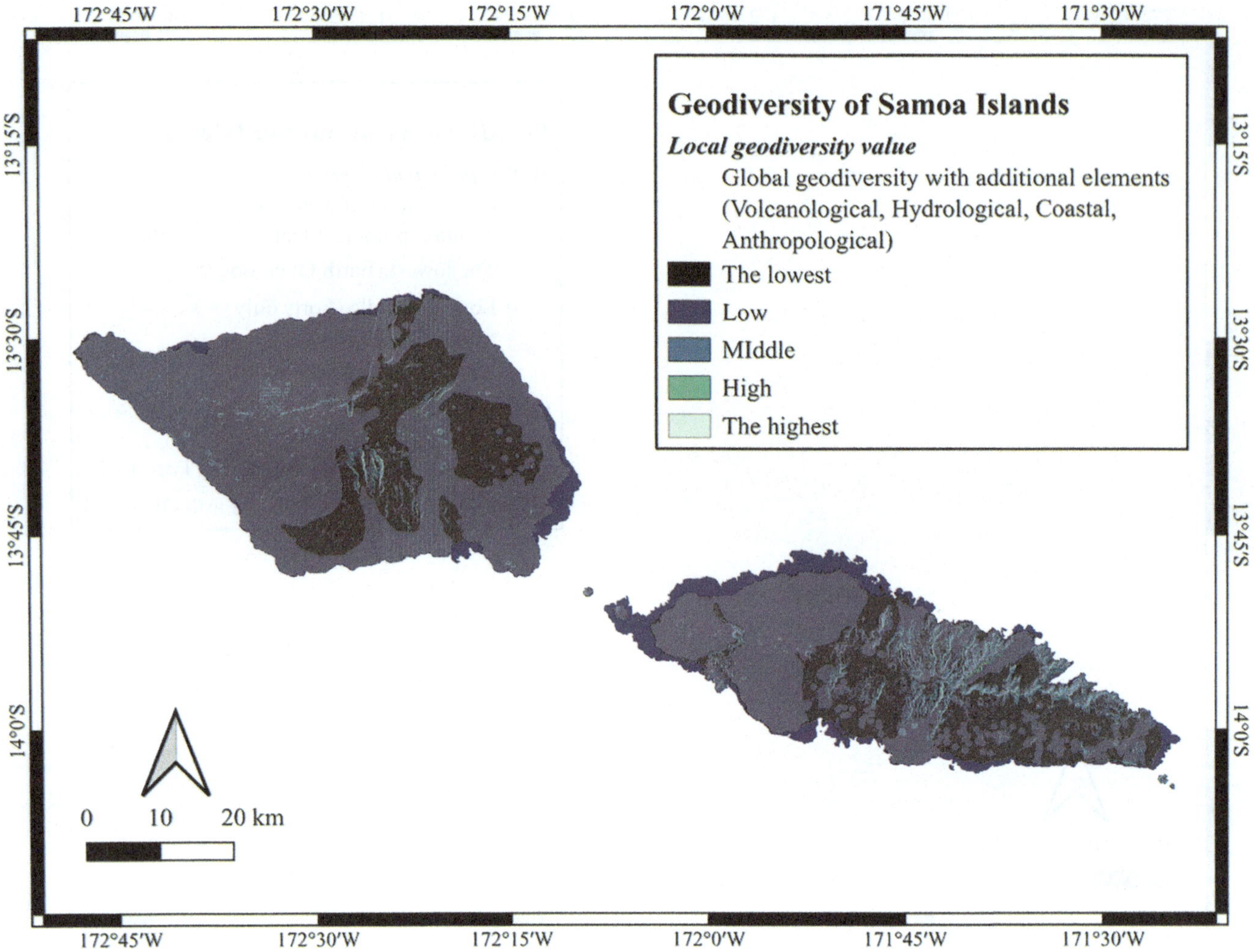

Fig. 3.28 Local geodiversity model based combined with the global geodiversity results show clearly where the areas geosites with high value for geoeducation, geotourism or geoconservation located in Samoa

relatively uniform morphology and geological features preventing to be able to refine the islands total geodiversity values. Samoa also provides a very good example to demonstrate the significance of local geodiversity that can be applied to the specific region to locate potential geosites that can form the basis of geoeducation, geotourism and geoconservation ventures (Fig. 3.28). Local geodiversity however is not suitable for global comparison. It can however be used to compare similar regions where the global assets are in the same range such as Samoa, to see the variations of diversity across similar geological and geomorphological attributes shown elsewhere.

References

Abbas MR, Hason MM, Ahmad BB, Abbas TR (2020) Surface roughness distribution map for Iraq using satellite data and GIS techniques. Arab J Geosci 13(17):839

Arifin YI, Pattiro WM, Manyoe IN, Napu SSS, Sugawara H (2023) Analysis and quantitative assessment of geodiversity at Karya Murni, Gorontalo, Indonesia. J Tour Geosites 48(2):763–773

Barančoková M, Hutárová D, Nikolaj M (2023) Quantitative assessment of geodiversity for conservation purposes in Slovenské rudohorie Mountains (Slovakia). Land 12(9):1650

Bétard F, Peulvast J-P (2019) Geodiversity hotspots: concept, method and cartographic application for geoconservation purposes at a regional scale. Environ Manag 63(6):822–834

Blatt H, Jones RL (1975) Proportions of exposed igneous, metamorphic, and sedimentary rocks. GSA Bull 86(8):1085–1088

Brilha J (2018) Geoheritage: inventories and evaluation. In: Geoheritage: assessment, protection, and management. Elsevier, Amsterdam, pp 69–85

Brilha J, Gray M, Pereira DI, Pereira P (2018) Geodiversity: an integrative review as a contribution to the sustainable management of the whole of nature. Environ Sci Policy 86:19–28

Burnelli M, Melelli L, Alvioli M (2023) Land surface diversity: a geomorphodiversity index of Italy. Earth Surf Process Landf 48(15):3025–3040

de Almeida Rangel L, do Carmo Oliveira Jorge M, Teixeira Guerra AJ, Fullen MA (2019) Geotourism and soil quality on trails within conservation units in South-East Brazil. Geoheritage 11(3):1151–1161

DeLong DC (1996) Defining biodiversity. Wildl Soc Bull 24(4):738–749
Devroye L, Kruszewski P (1996) On the Horton-Strahler number for random tries. RAIRO Theor Inf Appl 30(5):443–456
Dias MCSS, Domingos JO, dos Santos Costa SS, do Nascimento MAL, da Silva MLN, Granjeiro LP, de Lima Miranda RF (2021) Geodiversity index map of Rio Grande do Norte State, Northeast Brazil: cartography and quantitative assessment. Geoheritage 13:1
dos Santos FM, de La Corte BD, Saad AR, da Silva Ferreira AT (2020) Geodiversity index weighted by multivariate statistical analysis. Appl Geomat 12(3):361–370
Erfurt-Cooper P (2011) Geotourism in volcanic and geothermal environments: playing with fire? Geoheritage 3(3):187–193
Fepuleai A, Németh K (2019) Volcanic geoheritage of landslides and rockfalls on a tropical ocean island (Western Samoa, SW Pacific). Geoheritage 11(2):577–596
Forte JP, Brilha J, Pereira DI, Nolasco M (2018) Kernel density applied to the quantitative assessment of geodiversity. Geoheritage 10(2):205–217
Fossey M, Angers D, Bustany C, Cudennec C, Durand P, Gascuel-Odoux C, Jaffrezic A, Pérès G, Besse C, Walter C (2020) A framework to consider soil ecosystem services in territorial planning. Front Environ Sci 8:28
Habib M (2021) Quantifying topographic ruggedness using principal component analysis. Adv Civil Eng 2021(1):3311912
Jackmond G, Fonoti D, Meleisea M, Schoeffel P, Durling M, Tautunu MM, Sahib M (2022) UTU: Sāmoa archaeology and cultural heritage database. Archaeol Ocean 57(2):95–110
Jasiewicz J, Stepinski TF (2013) Geomorphons — a pattern recognition approach to classification and mapping of landforms. Geomorphology 182:147–156
Kear D, Wood BL (1959) The geology and hydrology of Western Samoa. N Z Geol Surv Bull 63:1–90
Keating BH (1991) Geology of the Samoan Islands. In: Keating Barbara H, Bolton Barrie R (eds) Geology and offshore mineral resources of the Central Pacific Basin. Circum-Pac Council Energy Mineral Resour, Houston, pp 127–178
Lausch A, Baade J, Bannehr L, Borg E, Bumberger J, Chabrilliat S, Dietrich P, Gerighausen H, Glässer C, Hacker JM, Haase D, Jagdhuber T, Jany S, Jung A, Karnieli A, Kraemer R, Makki M, Mielke C, Möller M, Mollenhauer H, Montzka C, Pause M, Rogass C, Rozenstein O, Schmullius C, Schrodt F, Schrön M, Schulz K, Schütze C, Schweitzer C, Selsam P, Skidmore AK, Spengler D, Thiel C, Truckenbrodt SC, Vohland M, Wagner R, Weber U, Werban U, Wollschläger U, Zacharias S, Schaepman ME (2019) Linking remote sensing and geodiversity and their traits relevant to biodiversity-part I: soil characteristics. Remote Sens 11(20):2356
Meten M, Bhandary NP, Yatabe R (2015) GIS-based frequency ratio and logistic regression modelling for landslide susceptibility mapping of Debre Sina area in central Ethiopia. J Mt Sci 12(6):1355–1372
Nazarenko I, Polchina S, Dmitruk J, Smaga I, Nikorich V (2006) Gruntoznavstvo z osnovami geologii. Chernivci Knigi 21:504
Quesada-Román A, Campos N, Alcalá-Reygosa J, Granados-Bolaños S (2020) Equilibrium-line altitude and temperature reconstructions during the Last Glacial Maximum in Chirripó National Park, Costa Rica. J S Am Earth Sci 100:102576
Quesada-Román A, Campos N, Granados-Bolaños S (2021) Tropical glacier reconstructions during the last glacial maximum in Costa Rica. Rev Mex Cien Geol 38(1):55–64
Reverte FC, Garcia MGM, Brilha J, Pellejero AU (2020) Assessment of impacts on ecosystem services provided by geodiversity in highly urbanised areas: a case study of the Taubaté Basin, Brazil. Environ Sci Pol 112:91–106
Riley SJ, De Gloria SD, Elliot R (1999) A terrain ruggedness index that quantifies topographic heterogeneity. Intermountain J Sci 5:23–27
Santos DS, Mansur KL, Seoane JCS, Mucivuna VC, Reynard E (2020) Methodological proposal for the inventory and assessment of geomorphosites: an integrated approach focused on territorial management and geoconservation. Environ Manag 66(3):476–497
Sarkar S (2002) Defining "biodiversity"; assessing biodiversity. Monist 85(1):131–155
Stepinski TF (2011) Geomorphons-a new approach to classification of landforms
Swingland IR (2013) Definition of biodiversity. In: Levin SA (ed) Encyclopedia of biodiversity, 2nd edn. Academic, Waltham, pp 399–410
Zaady E, Stavi I, Yizhaq H (2021) Hillslope geodiversity effects on properties and composition of biological soil crusts in drylands. Eur J Soil Sci 72(5):2308–2315
Zakharovskyi V, Németh K (2021) Qualitative-quantitative assessment of geodiversity of Western Samoa (SW Pacific) to identify places of interest for further geoconservation, geoeducation, and geotourism development. Geographies 1(3):362–380
Zakharovskyi V, Németh K (2022a) Geomorphological model comparison for geosites, utilizing qualitative–quantitative assessment of geodiversity, coromandel peninsula, New Zealand. Geographies 2(4):609–628
Zakharovskyi V, Németh K (2022b) Scale influence on qualitative–quantitative geodiversity assessments for the geosite recognition of Western Samoa. Geographies 2(3):476–490
Zakharovskyi V, Nemeth K, Gravis I, Twemlow C (2022) Geoheritage and geodiversity assessment framework for practical application to geoconservation of the coromandel Peninsula, New Zealand. Geoconserv Res 5(1):59–88
Zwoliński Z, Najwer A, Giardino M (2018) Methods for assessing geodiversity. In: Geoheritage: assessment, protection, and management. Elsevier, Amsterdam, pp 27–52

4 From Tourism to Geotourism

Tourism significantly impacts the global economy (Camilleri 2018), especially in small economies like island nations in the SW Pacific (Hall 1994; Hughes 2022; Kumar and Stauvermann 2023; Poirine 2010; Pratt and Harrison 2015; Rapaport 2013). In remote areas, tourism often leverages unique local features to attract visitors (D'Hauteserre 2016; Hughes 2022; Kreisel 1996; Kreisel and Reeh 2006). While it can boost local economies (Pratt and Kuilamu 2022), it may also strain local resources if poorly managed (Harrison 2022).

Niche tourism, targeting specific groups like medical or adventure tourists, plays a crucial role (Ali-Knight and Ensor 2017; Connell 2009; Elliot 2010; Hose 2004; Jennings 2007; Kuchumov et al. 2024; Mohanan and Shekhar 2022; Novelli 2004; Novelli and Benson 2004; Robinson and Novelli 2004). Conflicts between economic needs and local customs pose challenges, particularly in small islands with limited capacity for tourism (Buckwell et al. 2024; Lovelock et al. 2022). Sustainable tourism development is essential for balancing these factors in places like Samoa.

Geotourism is a form of niche tourism (Bull and Weed 1999; Hose 2008; Ólafsdóttir 2019) as it is based on the geological and geomorphological diversity and their appreciation through various activities providing direct experiencing of the lands (Dowling and Newsome 2005a, b; Hose 2004). Geotourism is a type of tourism that aims to sustain or enhance the geographical characteristics of a location (Dowling and Newsome 2005a, b; Newsome and Dowling 2005). Geotourism is defined as tourism that maintains or improves the geographical character of a place, including its environment, culture, aesthetics, heritage, and the well-being of its residents. It involves more than just "drive through" travel. Geotourism is a knowledge-based form of tourism that combines the tourism industry with the conservation and interpretation of abiotic nature features, while also considering related cultural issues within geosites for public engagement (Dowling 2011; Gray 2008).

Geotourism, while initially defined as the study of geological and geomorphological features, now encompasses a broader concept within abiotic nature-based tourism. It is knowledge-based tourism that integrates the promotion and conservation of geological features into tourism activities. This involves identifying key geosites and creating frameworks to share geoscience knowledge with the public. Geotourism specifically focuses on landscape and geology within natural area tourism (Dowling and Newsome 2010). It promotes tourism to geosites, conserves geo-diversity, and enhances understanding of Earth sciences through appreciation and learning (Dowling and Newsome 2010). This is accomplished through independent visits to geological features, using geo-trails and viewpoints, guided tours, geo-activities, and visiting geosite visitor centers (Dowling and Newsome 2010). It is important to note the differences between definitions, highlighting whether they are based on independent or oriented tourism activities. Additionally, there are various definitions that emphasize the educational and science-based approach, focusing on the geoeducation aspect (Hose 2012). This framework helps clarify the distinction between geotourism and ecotourism. Although they can overlap, geotourism focuses on the abiotic nature, unlike ecotourism, which is biologically centered. Geotourism emphasizes geodiversity and geoheritage, focusing on three key aspects: (1) Abiotic nature as the main attraction, (2) Geological heritage interpretation, and (3) Strong geoconservation efforts benefiting both nature and visitors.

In summary, geotourism focuses on the geodiversity and geoheritage of a region, highlighting sites that appeal to visitors for geoeducation and visual interest related to the abiotic nature.

A geosite is a significant location within the concept of geodiversity, notable for its geological or geomorphological features. While primarily identified by these characteristics, it can also have cultural or heritage importance. Various methods have emerged over the past two decades to refine

K. Németh, V. Zakharovskyi, *Geodiversity of Samoa*, Geoheritage, Geoparks and Geotourism,
https://doi.org/10.1007/978-3-031-87220-4_4

geosite selection criteria. Some experts focus strictly on geological and geomorphological attributes, while others include geocultural aspects. Geosites, evaluated through audits, assessments, and selection processes, face various threats (Prosser 2018). Geosite recognition can utilize qualitative, quantitative, or combined methodologies (Berrezueta et al. 2021; Carrión-Mero et al. 2024a, b; Cvetanović et al. 2024; Dim 2024; Golfinopoulos et al. 2022; Guilbaud et al. 2024; La Russa et al. 2024; Migoń 2024; Pimentel et al. 2024; Quesada-Valverde and Quesada-Román 2025; Santos and Mansur 2024; Somma et al. 2024; Spyrou et al. 2024; Tamang et al. 2023; Zakharovskyi et al. 2023, 2024). These methods all function by recognizing representativeness, rarity, and certain functionalities to define a place as a geosite (Brilha 2016; Cameron 2024; Cleal et al. 1999; Diniz and de Araújo 2022; Herrera-Franco et al. 2020; Tomić and Božić 2014; Vujičić et al. 2011). Geosite recognition often lead to some sort of design of geotrails or georoutes (Gravis et al. 2020; Mahboubi and Naimi 2024) involving complex communication link to the public (Pasquaré Mariotto et al. 2023).

Geoheritage conservation is an essential aspect of geotourism development because it ensures the long-term sustainability and preservation of geological and geomorphological values (Ferdowsi 2024).

The development of geotourism must be integrated with other tourism activities in the region, ensuring interconnectivity. This requires proper identification of the area's geodiversity and geoheritage. To achieve sustainable geotourism, it is essential to thoroughly understand the region's tourism structure, including visitor types, their needs, and spending patterns (Ali-Knight and Ensor 2017; Constantin et al. 2021; Kelly 2012; Valentine et al. 2004). Geotourism requires a complete social science approach and ongoing monitoring to fine-tune various aspects. It must align with an effective geoconservation strategy within the broader conservation framework. Geotourism can be done independently or through guided tours. The dominant form depends on natural conditions, community structure, land ownership, region's capacity, development level, and potential hazards. A balanced approach allows for both individual travelers and group tours. Guided tours offer social engagement and enhance visitor experiences (Cheung 2016; El-sharkawy 2015; Frey et al. 2021).

Small island nations face challenges in geotourism development due to their size, indigenous cultures, complex ecosystems, limited capacity, and varied visitor traits (Németh and Gravis 2022). Geotourism in small island nations faces challenges due to limited resources and lack of interest from governments and local leaders. In customary lands, communities' strong ties to the land often oppose tourism. Consequently, even areas with notable geosites and high geodiversity may be excluded from geotourism development.

In small island nations, geopark programs have emerged as a potential means for promoting geotourism. However, the differing visions between developers and local communities often impede the progress of exploring geotourism based on mutual benefits (Fepuleai et al. 2021; Németh and Gravis 2022).

The next sections will summarize Samoa's major tourism structure and its potential for sustainable geotourism development.

4.1 Samoa Tourism

Samoa tourism developed significantly since the year 2000. Samoa increasingly became a popular tropical holiday destination (Pearce 2002). Samoa's tourism marketing strategy is primarily focused on its natural beauty, tropical appearance, and pristine beaches (Fig. 4.1). Upon arrival at Falelolo International Airport (Fig. 4.2), visitors are immediately enveloped in the serene Pacific atmosphere. This ambiance is consistent throughout the Pacific region, often depicted as a place to escape worldly worries, particularly from a Western perspective.

Tourism information and services are readily available upon arrival, allowing independent travelers to navigate with ease. Over the past two decades, Samoa has become an adept host nation, effectively catering to visitor needs in a straightforward and efficient manner. Samoa remains a key destination for large group tours, including cruise ship visits. Large resort-style hotels predominantly serve family groups and operate as self-contained entities, providing comprehensive services including organized excursions.

Furthermore, small businesses are prevalent in Samoa, facilitating easy access for independent visitors to major centers and significant beach sites even on a modest budget. In summary, Samoa boasts a diverse and well-developed tourism sector capable of accommodating both budget-conscious and luxury travelers.

In Samoa, as well as in other small Polynesian islands, it is customary for most tourism accommodations to be locally owned and operated.

While many Southwest Pacific islands promote sun, sand, and surf tourism, Samoa's tourism is distinctive for its unique development (Kiddell 2012). Over 80% of Samoan land is customarily owned, limiting international tourism investment. Thus, unlike Fiji, Tonga, and the Cook Islands, most of Samoa's accommodations are locally owned and operated (Kiddell 2012). Customary land ownership also enables villagers to share profits. For example, locals collect small fees charged to tourists for accessing natural sites such as beaches and waterfalls. This system helps distribute wealth and addresses the common trend where tourism development often benefits the wealthier

Fig. 4.1 A typical vision of Samoa as a Pacific Island where pristine coral sand beaches and coconut fringed coasts are the main attraction for visitation (14°02′25″S, 171°29′43″W)

residents of a host community more than the poorer inhabitants (Harrison 2008).

Family-run tourism businesses in Samoa often follow customary and village rules, benefiting extended families economically. Despite the rise of foreign-owned large tourism ventures, low-budget accommodations like thatched beach fale huts continue to emerge along the coast (Schänzel et al. 2022). These beach fale, once popular with backpackers, are now increasingly attracting visiting families from overseas (Schänzel et al. 2022). Slow tourism, though rarely fully studied, is popular among budget travelers and those seeking local culture, culinary experiences, and a break from fast-paced life. This trend benefits local businesses and attracts global visitors, promising sustainable tourism growth in Samoa.

Beside rental cars or taxis, Samoa has an effective but slow public bus network that provides a unique experience to the visitors (Fig. 4.3). The Samoan buses are part of the local culture and operated by villages (Fig. 4.4). Long distance travel can be time consuming for the visitors, but this form of travel offers direct opportunity to feel the Samoan time.

The natural beauty of the island alongside with its rich Polynesian culture offer cultural experiences to the visitors such as the common dance night that no visitors can miss as they appear on many occasions from village ceremonies to luxury resort programs (Fig. 4.5). As dance and tattooing has deep cultural meaning and symbology many linked to the geoheritage of the islands the visitor can't miss to see and experience this rich cultural vibration in Samoa (Fepuleai et al. 2017).

The tropical atmosphere just makes an even more relax atmosphere as all year long flowering of the forest and the tidy garden like appearance of villages and plantations provide a unique backdrop of the colorful Samoan environment (Figs. 4.6 and 4.7). The Samoan islands smell nearly all year long amazing dominated by the sweet scent of moso'oi trees (Cananga odorata, known also as ylang-ylang) blooming (Fig. 4.6). The islands bird life also fascinating, and the abundant flowering trees can attract tropical birds such as the moso'oi tree fruits, that attracts native birds including the Pacific Pigeon (Ducula pacifica), informally called as "lupe" or the purple-capped-fruit dove ("manutagi" as Ptilinopus ponapensis) who love to eat the fruit.

The decision to choose Samoa as a travel destination is primarily influenced by three factors: (1) the presence of family or friends in Samoa, (2) the appeal of beautiful beaches, warm water, and scenic views, and (3) favorable warm and sunny weather. Tourists from Europe additionally

Fig. 4.2 Faleolo International Airport in NW Upolu as a main international gateway to Samoa (13°49′56″S, 171°59′49″W)

value the cultural and historical aspects, albeit to a lesser extent than family connections. Holiday tourists are mainly drawn to the beautiful beaches, warm waters, and sunny weather.

The most common sources of information for trip planning include word-of-mouth recommendations from friends and relatives. Approximately 23% of tourists rely on previous experiences; however, this is less applicable to European visitors due to the predominance of first-time visits from this group. Consequently, European visitors tend to utilize web resources, books, and other published media for their travel planning.

Data shows most visitors plan their trips based on personal recommendations, while European visitors rely more on electronic or printed media, though they make up a small percentage of all visitors.

While tourism in respect to Samoa normally considering inbound visitation by foreigners, there is a considerable deep tradition of tourism from Samoa and within Samoa by Samoans. This is far less studied subject hence very limited information is available about its structure, style or needs. Samoan understandings of travel and tourism are expressed via the concept of *malaga*, which has a range of meanings including '*migration*', '*movement*' or '*travel back and forth*' (Gibson et al. 2022; Lilomaiava-Doktor 2009; Raymond 2023). For Samoans, travel has historically been driven by cultural and familial purposes, including the maintenance of cultural bonds through fa'alavelave (traditional obligations)

Fig. 4.3 Colorful busses of Samoa (here in Savai'i) are vital part of the visual appearance of Samoa

Fig. 4.4 Public transportations of Samoa (here in Upolu) are a real cultural experience. Despite they are slow services, nearly every corner of both islands accessible by village operated bus services (weekends frequency can be very low)

and visiting friends and relatives (VFR). The primary internal travel patterns of locals reflect these concepts, which differ significantly from the visitation patterns of non-Samoans to the island, necessitating distinct catering approaches.

Additionally, economic migration within the Pacific, particularly to Australia, New Zealand, and the USA, has increased the need for travel among Samoans, even among older generations. This migration has restructured Samoan society into a more mobile community compared to previous decades. Both urban and rural Samoans' travel mobilities are influenced by a combination of cultural and familial obligations, leisure and tourism goals, and work-related purposes, thereby blurring the lines between tourism and other forms of mobility.

Samoan tourism often involves purposeful movement, with leisure being a component rather than the primary objective, unlike the predominant focus of non-Samoan international visitors. Consequently, finding space for geotourism in this context is a complex and unexplored area. However, it presents a potential subject for future research, given the increasing geopolitical pressures in the South-West Pacific, economic volatility, and proximity to New Zealand and Australia. These factors naturally bring about a new generation of travelers with a growing emphasis on various aspects of rejuvenation. This new generation possesses the financial freedom and spending power to explore activities such as geotourism, which were previously inaccessible or overlooked.

4.1.1 Who Visits Samoa?

Since 2014, the Samoa Tourism Authority (https://www.samoatourism.org/. Accessed 18 December 2024) has systematically gathered extensive visitor data. Initially, visitors provided basic information on an arrival card; however, this process has since evolved into a more comprehensive data collection system. The COVID-19 pandemic led to a lockdown in Samoa, which disrupted data collection efforts and resulted in zero visitor entries during that period. Despite these challenges, the available data still reveals important trends in Samoa's tourism dynamics.

From 2014 to 2019, there was minimal change in the visitation structure. Despite minor fluctuations in visitor numbers, the primary composition of visitors remained relatively stable. Visitors from New Zealand constituted 44% of all holiday arrivals, although only slightly over one-third of all arrivals from New Zealand were for holiday purposes. This

Fig. 4.5 Dance shows are part of tourism programs in every level and can be a great cultural experience. While the shows provide visual attraction, full explanation of the symbology of the dance moves are rare provided. Image is from the Tanu Beach Fales, Savai'i

Fig. 4.6 Samoa tropical climate provides a rich vegetation with exotic flowers. While most of the plants are part of traditional medicine and cultural symbology, sweet scent flowers such as moso'oi (Cananga odorata, known as ylang-ylang), commonly part of traditional welcome flower necklaces that are produced by village communities and delivered to main tourist sites and airports

may be indicative of the business connections between Samoa and New Zealand, as well as the significant number of returning Samoan citizens and their families residing in New Zealand. European visitors made up approximately 7% of the total, with 73% traveling for holiday purposes (Figs. 4.8, 4.9, 4.10, 4.11, 4.12, and 4.13). Australians represented the second largest group of visitors, displaying a purpose structure like New Zealanders. Business connections and family visits likely contributed to the non-holiday trips for Australians. Overall, one-quarter of all holiday tourists to Samoa originated from Australia (Figs. 4.8, 4.9, 4.10, 4.11, 4.12, and 4.13).

Nearly 50% of all visitors from American Samoa were Samoans, with 43% being United States nationals. Analysis showed that over half (54.9%) of all tourist arrivals to Samoa had visited the island before. The majority (93%) of visitors from American Samoa were returning visitors, likely of Samoan origin. In contrast, 94% of Europeans were first-time visitors to Samoa. Despite expectations of significant Asian visitation, statistics up to 2019 did not clearly show this trend. This might be due to the challenging and expensive access routes from Asia to Samoa. However, it is projected that the presence of Chinese interest and business enterprises will lead to an increase in visitation from China in recent years.

The impact of COVID-19 on Samoa's life (Aumua et al. 2023; Connell 2021; Kumar and Stauvermann 2023) including in its tourism sector is evident in the newly released visitation statistics (https://www.samoatourism.org/. Accessed 18 December 2024) (Figs. 4.14, 4.15, and 4.16). Despite the dramatic impact Samoans showed a remarkable resilience to support their home in every possible ways that helped to come over the difficult times and be able to reignite their community and economy (Dé and Jackson-Becerra 2021). The most notable change in the tourism since the pandemic it that besides the overall decline in total visitation numbers, is that visitors from countries other than New Zealand and Australia have not returned to Samoa in the same numbers as they did before the pandemic. The predominance of visitors from New Zealand and Australia suggests that tourism ventures should be more focused on catering to their needs and developing niche tourism programs aligned with their inter-

Fig. 4.7 Fagraea berteroana, commonly known as the pua keni keni, pua kenikeni or perfume flower tree, short is just pua flower as a symbolic flowering shrub can be seen and smell everywhere in Samoa. It is a symbolic flower for many art form and village decorations, a truly exotic feature of Samoa, providing great visual appearance to attract visitors

ests. Additionally, targeted marketing campaigns may be necessary to regain tourists from Europe and Asia to pre-COVID-19 levels.

COVID-19 is clearly isolated SW Pacific Island states from international tourism (Kumar and Stauvermann 2023; Movono et al. 2022). Many of these states were/are heavily dependent on the inbound visitation as international tourism is a primary provider of employment and driver of economic development for most of the Pacific region. The lack of appropriate level of funding by most Pacific state governments clearly disrupted the tourism sector, interestingly despite of the loss of jobs and expected economic downturn the local communities were able to go through this time without significant impact on their well-being in both spiritual and even financial aspects, showing a remarkable resilience of the Pacific societies (Scheyvens et al. 2023). Research demonstrated that while communities faced financial struggles, their mental, social, physical, spiritual and environmental wellbeing had improved in many cases showing a very effective adaptivity to the unusual situation strongly based on the strong and intact community spirit the Pacific Islands, such as Samoa demonstrated (Scheyvens et al. 2023).

4.1.2 Tourists' Profile

The tourist profile of Samoan visitors (Fig. 4.17) is a crucial factor to consider for future tourism developments. Tourist profile can identify key fields where development is needed or areas from where the tourism functioning well (Ali-Knight

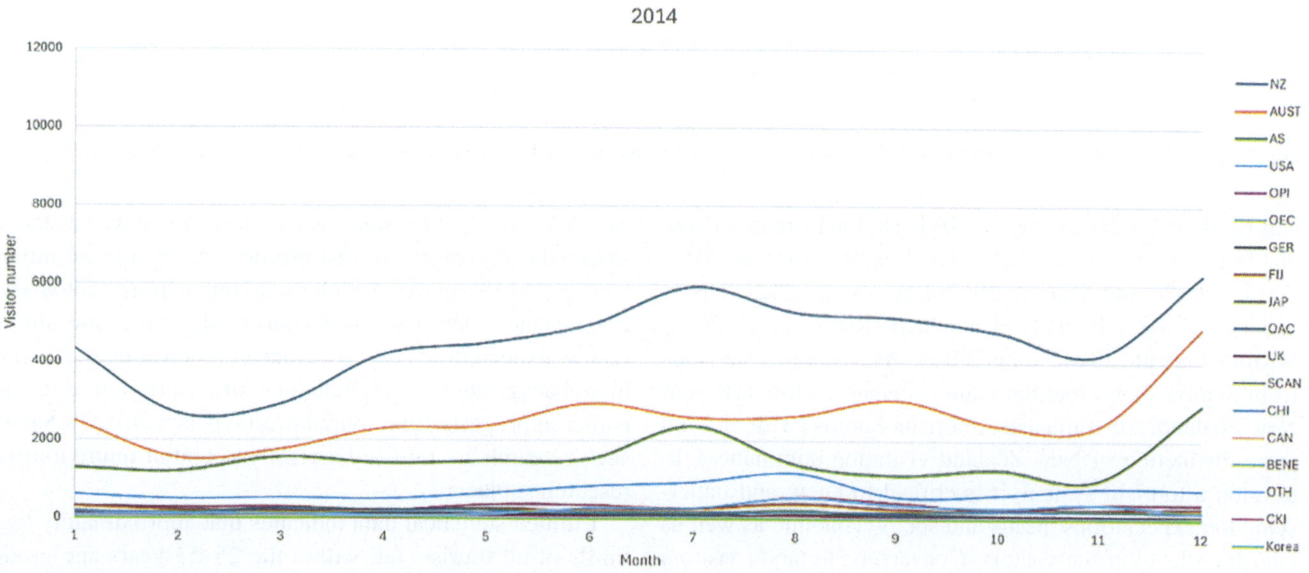

Fig. 4.8 Visitation structure of Samoa in 2014 based on the data set of the Samoa Tourism Authority (https://www.samoatourism.org/)

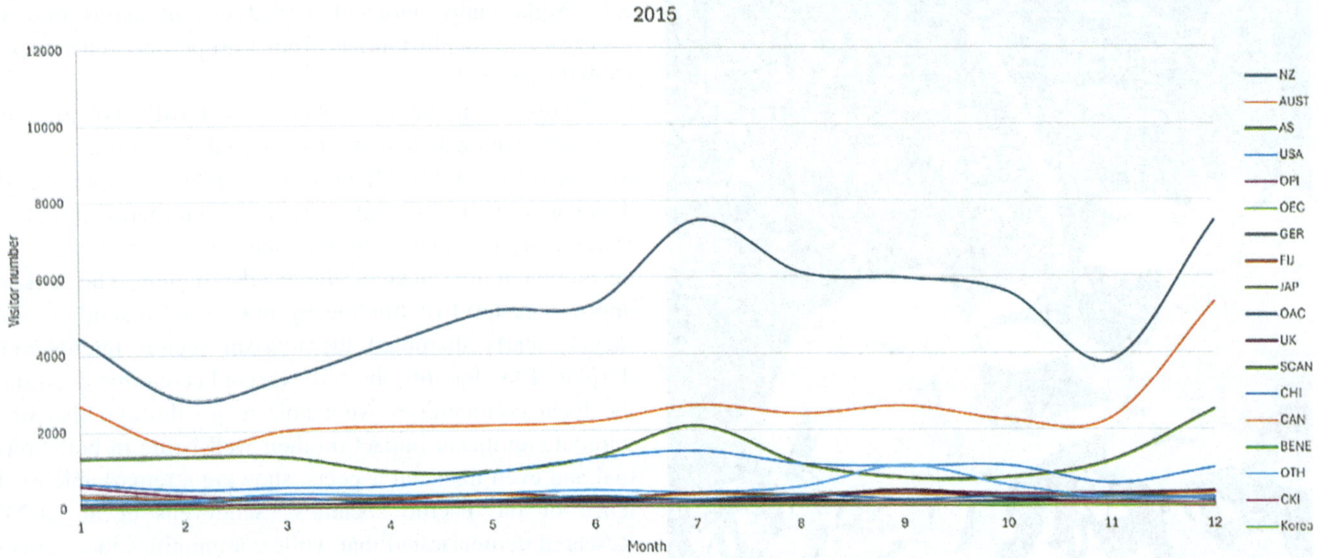

Fig. 4.9 Visitation structure of Samoa in 2015 based on the data set of the Samoa Tourism Authority (https://www.samoatourism.org/)

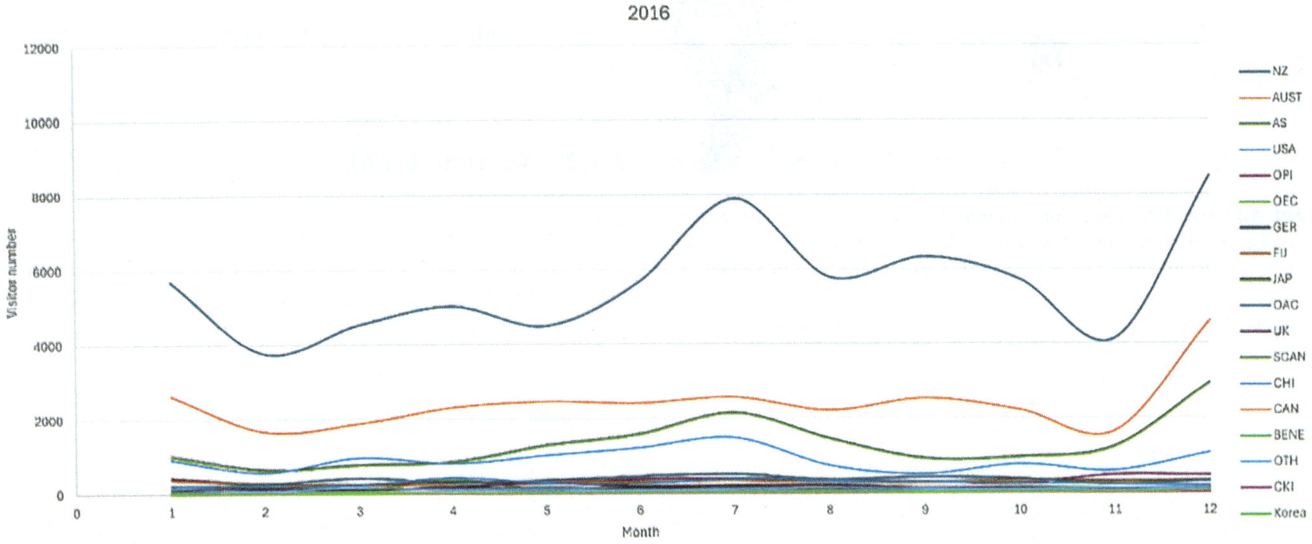

Fig. 4.10 Visitation structure of Samoa in 2016 based on the data set of the Samoa Tourism Authority (https://www.samoatourism.org/)

and Ensor 2017; Barutçu et al. 2011; De La Torre and Pérez 2017; Do Santos et al. 2021; Fonseca Filho et al. 2018; Fonseca Filho and Moreira 2017; Lopes et al. 2024; Nistor and Dezsi 2022; Piedade et al. 2024; Rivero et al. 2019; Vasiljević et al. 2018; Vong 2016). As visitation statistics from Samoa shows that the main overseas visitors are from New Zealand, Australia and American Samoa (with an even more distinction of New Zealand visitation dominance), to develop a tourism system, it is critical to know and understand the expectations, needs and socio-economic as well as education level of the visitors. Conversely, to target visitors with potential higher consumer power (e.g., European and North American), the good knowledge of their "background" can help to identify areas where investment is needed to attract those visitors. Tourist profiles are even more important to find the place of niche tourism such as geotourism. For instance, one tourism design maybe either too simple (below expectations) or too advanced (not fitting to the common background or expectations of visitors) if it is not explored properly from tourist profile perspective. In Samoa such research are rare and certainly potential future tourism research subjects.

Limited statistical data indicates that approximately two-thirds of all tourists fall within the 25–54 years age group, though there are significant differences among various generating countries. Notably, Samoa attracts potentially

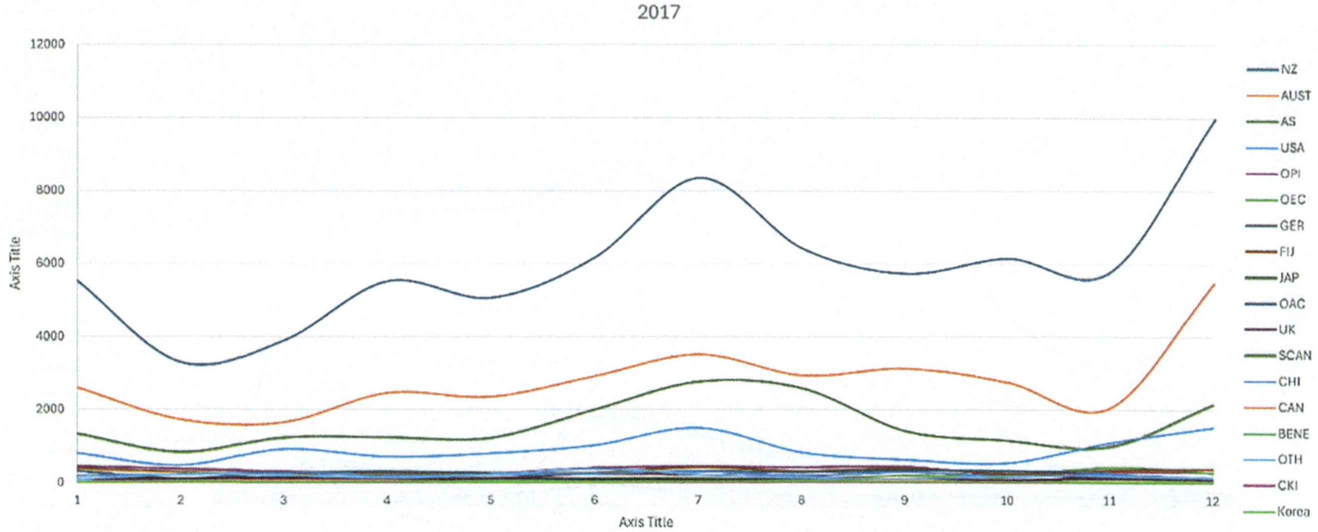

Fig. 4.11 Visitation structure of Samoa in 2017 based on the data set of the Samoa Tourism Authority (https://www.samoatourism.org/)

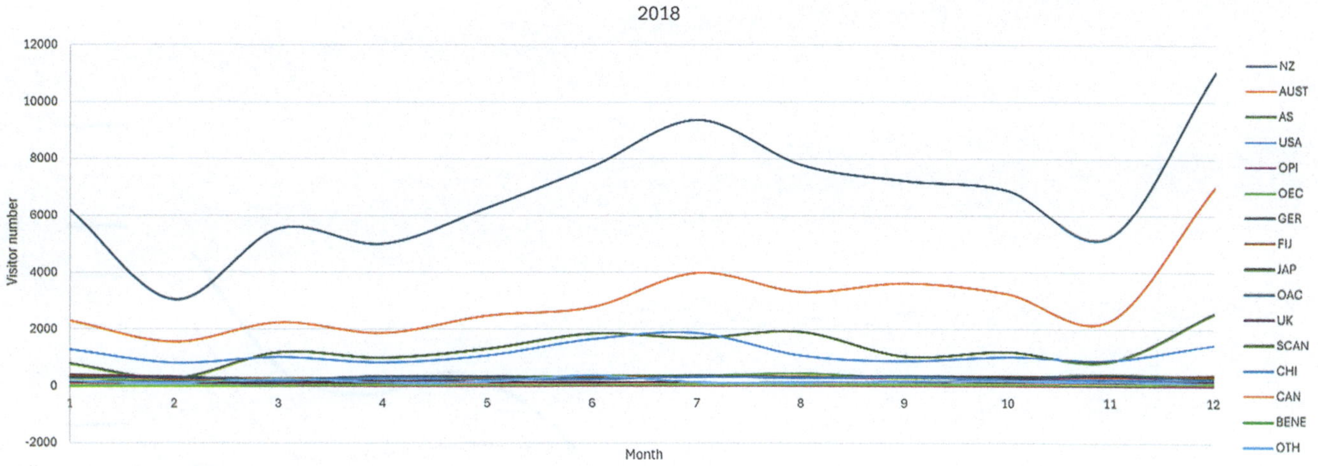

Fig. 4.12 Visitation structure of Samoa in 2018 based on the data set of the Samoa Tourism Authority (https://www.samoatourism.org/)

wealthy North Americans primarily in the 50–59 age group, whereas most European visitors are younger, predominantly within the 20–34 age group. The largest visitor group from New Zealand shows a relatively even age distribution, while Australian visitors clearly divide into young and late mid-age groups.

When comparing visitor ages to the main purposes of their trips, the holiday category is represented by the youngest age group. In terms of gender distribution, female travelers constitute 56% of visitors, with the highest percentages arriving from American Samoa (61%) and New Zealand (59%). It is also evident from the few years of data available that nearly 50% of holiday tourists travel with their spouse or partner. Conversely, business travelers are almost exclusively solo.

4.1.3 Tourism Ventures

Since around 2000, Samoa has seen significant growth in tourism, becoming a major destination in the SW Pacific for various budgets (Kiddell 2012; Kumar and Stauvermann 2023). The rapid expansion is evident through efforts to support business development (Samoa Tourism Authority. Accessed 17 December 2024). In 2024, the Samoan Tourism Authority provides accessible information about the sector online (https://www.samoatourism.org/. Accessed 17 December 2024). This summary highlights aspects of Samoan tourism related to geotourism, geoeducation, and geoconservation, preserving the island's diversity. Attractions have evolved over recent decades, and by 2024, tour operators clearly structure their programs around these sites. A

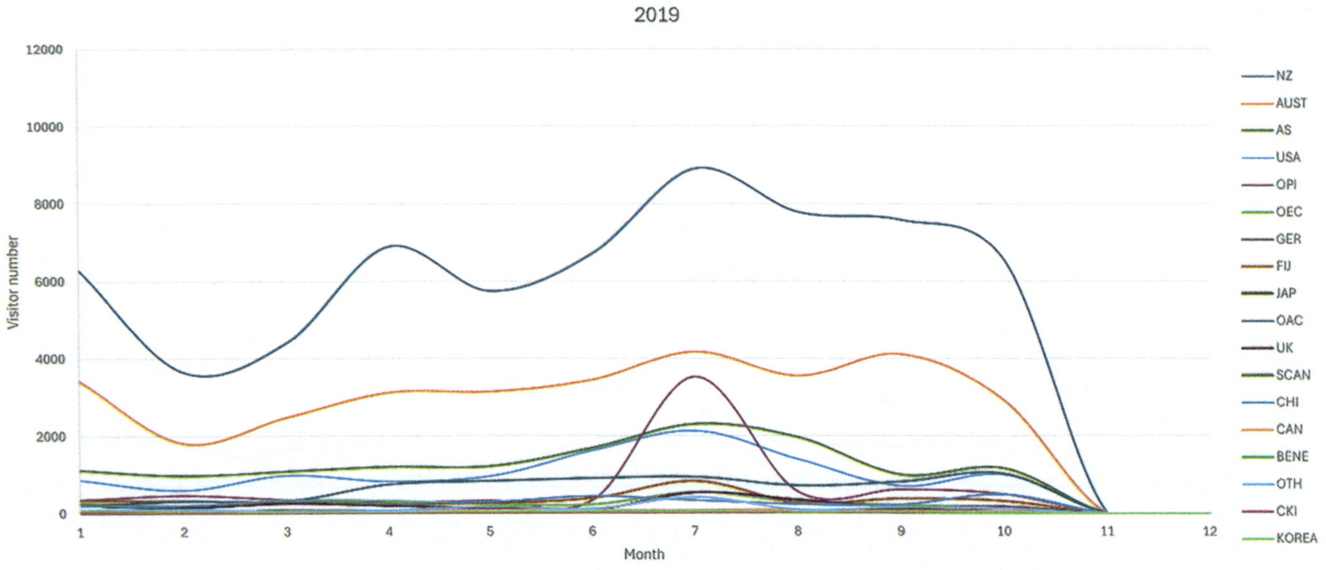

Fig. 4.13 Visitation structure of Samoa in 2019 based on the data set of the Samoa Tourism Authority (https://www.samoatourism.org/)

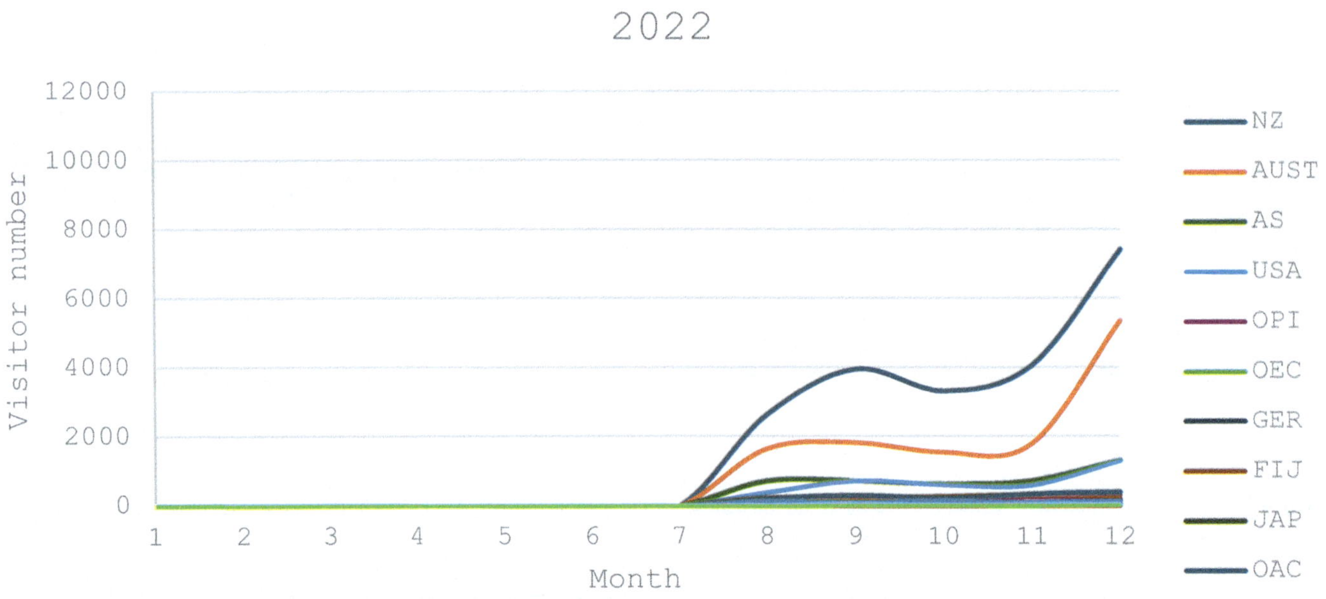

Fig. 4.14 Slow recovery on visitation numbers following the opening of the borders after the COVID-19 pandemic

downloadable map is available for private or business use (Fig. 4.18).

This trend reflects that especially families seeking more authentic, immersive and socially interactive experiences on holiday that are more akin to the local (Samoan) way of life or Fa'asamoa. It emerged that beach fale (Fig. 4.19) tourism provides culturally, culinary, socially and sensually richer experiences as part of slow hospitality offerings for families. The-family perspectives on holiday, slow hospitality, the beach fale accommodation developments, seaside (Agarwal 1997, 2002; Battilani 2024; Blumrodt and Palmer 2014; Bouvet et al. 2001) and/or thalasso tourism (Gomes et al. 2021; Lee and Lee 2015; Rodrigues et al. 2022) perfectly aligned with the general Samoan societal architecture hence can be considered as a sustainable economic model culturally deeply embedded to the Polynesian traditions (Schänzel et al. 2022). These days, beach fale "network" developed to be as a business model and referred to an as "industry" within tourism in Samoa contributing to the village economy significantly (Scheyvens 2005). Studies shows the sustainability of this tourism mode as they use local materials to

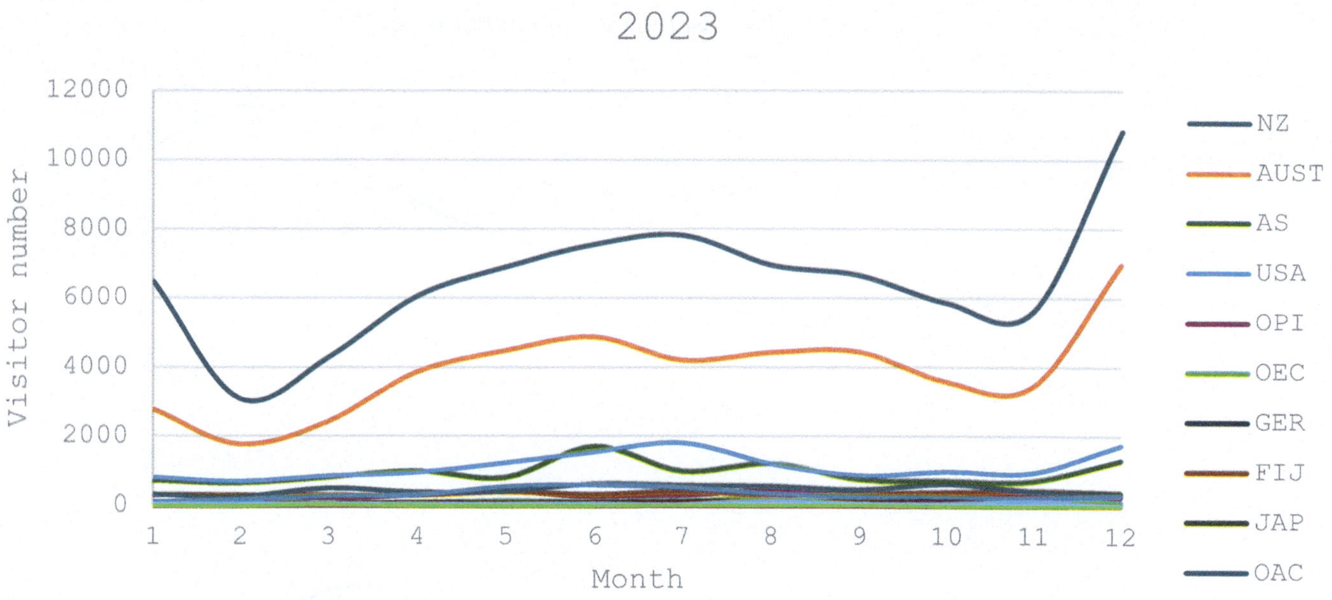

Fig. 4.15 Visitation is in steady growth in 2023 but the level of visitor numbers from other than New Zealand and Australia, stayed low across 2023

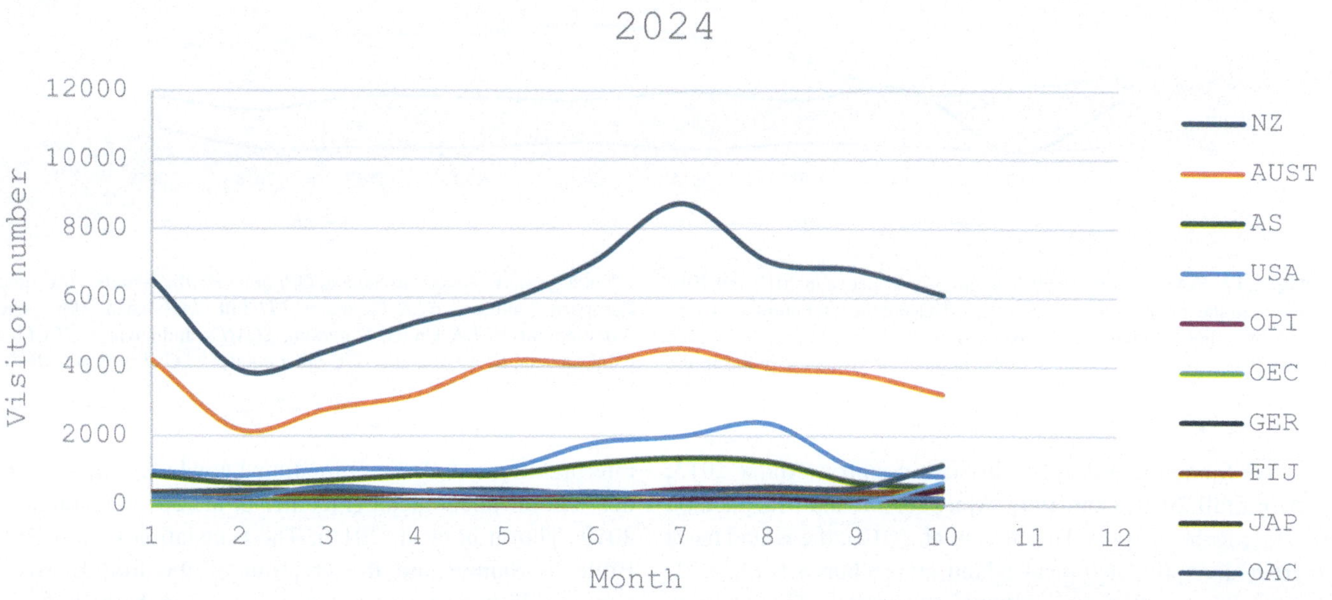

Fig. 4.16 In 2024 slight increase can be recognized in New Zealand visitor numbers. Visitors from USA probably consists of some cruse visitors. Other visitations from distant areas still in lower-than-expected levels

build, local workforce to manage and have income directly contributing the local community (Scheyvens 2005). Interestingly, despite effort to promote beach fale tourism, their visitor nights or as a resource largely invisible from tourism statistics (Scheyvens 2005).

Samoa offers great variety of experiences that can provide a full immersion to the vibrant Polynesian tropical life including various market visits (Figs. 4.20, 4.21, and 4.22). Market visits are also commonly organized for groups or guest of mid to upper class hotels or resorts but can be a highlight for any independent tourist. In the market places the visitor can explore local arts (Fig. 4.20), produces that are heavily influenced by seasonality (Fig. 4.21). More adventure seeking activity to check the local fish markets where the diversity of the Pacific marine life is on real life display (Fig. 4.22).

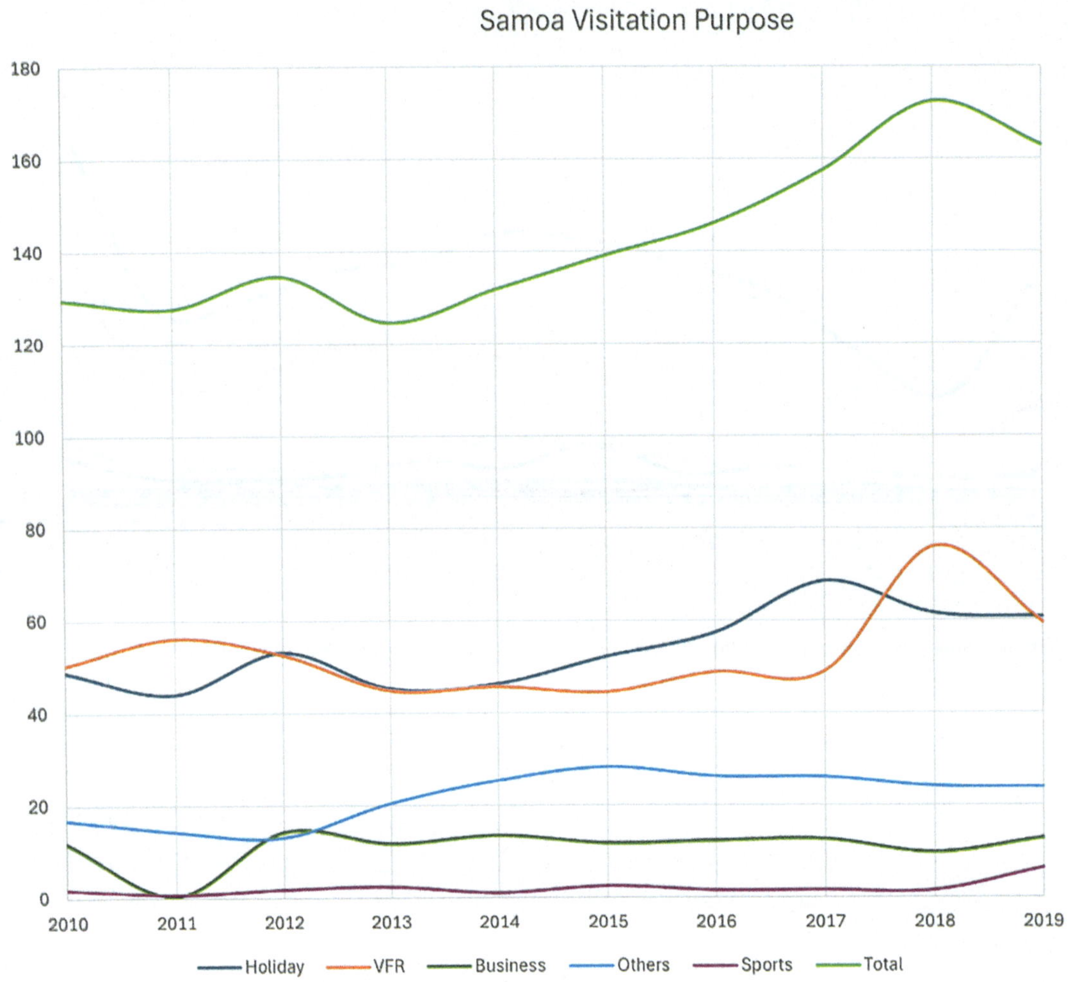

Fig. 4.17 Total visitor arrivals by country/year between 2010 and 2019 sorted in the main purpose of their trips. This are the total number of visitors by country arriving in Samoa by month. The total number for each month is shown along with the total number of visitors arriving in the calendar year. *AS* American Samoa, *OPI* other Pacific islands, *OEC* other European countries, *GER* Germany, *FIJ* Fiji, *JAP* Japan, *OAC* other Asian countries, *UK* United Kingdom, *SCAN* scandinavia, *CHI* China, *CAN* Canada, *BENE* Benelux, *OC* other countries, *CKI* cook islands

As Samoa is a deeply Christian country (Ahdar 2013; Alofaituli 2023; Côté 1997; Macpherson and Macpherson 2011; Robson 2009; Thornton et al. 2010), the island has a great diversity of typical island style church (Fig. 4.23) providing a distinct cultural experience for visitors. Religion is also an important part of various geodisaster interpretations and disseminating knowledge about various natural phenomena (Fletcher et al. 2013; Holmgaard 2019; Thornton et al. 2012). The translation of the Holy Bible to Samoa and the creation of the first English–Samoan dictionary created the literacy in Samoa that is also commemorated in stones in the church in Savai'i it is

Fig. 4.18 Feature map that can be accessed freely and distributed extensively across the island's main visitation sites showing the main roading and key tourist sites generally considered to be destinations for independent visitors (https://specialist.samoa.travel/home/page/ 16–18 December 2024)

considered to happen (Fig. 4.24). Locations with such cultural importance often linked to some features that can be connected to some level to geoheritage if nothing else the location itself (Fig. 4.24).

While most of the tour operators offer some firsthand experience of Samoan everyday life such as preparation of umu (rock oven) (Fig. 4.25 or Koko Samoa (Samoan chocolate drink) (Fig. 4.26) most of such programs are very artificial and serving events within broader cultural presentations. There is still an apparent lack of an effective way how independent foreign visitors can be engaged in such events in a genuine village visit style. This is probably heavily affected by the commonly strict village culture, restrictions and obligations (Ta'isi Efi 2018).

Fig. 4.19 Traditional local community operated beach fales (Matareva Beach) that can be rented for day or night visits, or just as a base for beach activities (13°59′21″S, 171°55′53″W)

Fig. 4.20 Samoa's capital, Apia has a large marketplace where traditional arts and their popular version can be accessed- The marketplace function as a tourist hotspot for visitors (13°50′03″S, 171°46′18″W)

Fig. 4.21 Apia market is also a good location to explore the island produces and enjoy the tropical variety of goods (13°50′03″S, 171°46′18″W)

Fig. 4.22 Local fish markets in Apia is a good tourist visitation area that best to visit in early mornings to see the richness of the SW Pacific marine environment that local communities utilize for their living greatly. The place is not a common tourist destination despite it is an interesting location to experience local living Polynesian culture (13°49′46″S, 171°46′12″W)

Fig. 4.23 As Samoa has a deep religious Christian root the islands are abundant in churches that are visually attractive and well maintained. In Sundays the church services visited by locals that are interesting experiences for visitors to see. In this image the Immaculate Conception Cathedral, in Apia is shown (13°49′55″S, 171°45′59″W)

Fig. 4.24 The memorial stone for the first Samoan translation of the Holy Bible in Avao Congregational Christian Church in Savai'i (13°26′55″S, 172°21′42″W). The translation was completed by George Pratt (1817–1894) who was a missionary with the London Missionary Society who lived in Samoa for 40 years from 1839 to 1879. He also created the first English–Samoan Dictionary (Pratt, George 1862). A Samoan Dictionary: English and Samoan, and Samoan and English; with a Short Grammar of the Samoan Dialect (first ed.). Samoa: London Missionary Society's Press)

Fig. 4.25 Traditional umu with pork prepared for special occasions by using heated stones (https://www.samoa.travel/experience/foodies/umu/. Accessed 18 December 2024)

Fig. 4.26 Family roasting of the seed of cocoa fruits to make Koko Samoa or Samoan chocolate drink (http://ultimatehotchocolate.com/2022/09/02/all-about-koko-samoa-samoas-national-hot-chocolate/. Accessed 18 December 2024)

4.2 Place of Geotourism

Through the Samoan Tourism Authority, a well-designed list of attraction sites with downloadable explanatory panels is available freely (https://www.samoatourism.org/section/64/attraction-sites. Accessed 14 December 2024). The selection of attraction sites aligns with national conservation goals, including geodiversity. Many key attraction sites feature significant geological characteristics which are promoted for their "geofeatures". The Samoan Tourism Authority defines Samoa as a potential adventure tourism destination (Fig. 4.27) due to its natural geodiversity and associated activities such as hiking, climbing, biking, and various water-based activities. This approach highlights the role of geotourism within Samoan tourism. Samoa represents an opportunity for further development in geotourism. Proper documentation of geological and geomorphological sites, along with geocultural locations, suggests significant potential for investment in geotourism. Investing resources into this area could potentially increase visitor numbers and spending, offering opportunities to explore the island's geoheritage in more depth. Additionally, given Samoa's exposure to diverse geohazards, promoting geoheritage through geotourism could support natural hazard resilience education (Avagyan et al. 2023; Kubalíková et al. 2023; Mehdipour Ghazi and Audra 2022; Németh 2022; Németh et al. 2021a, b; Novković et al. 2022; Vandelli et al. 2024; Wang et al. 2019).

Samoa's tropical conditions and abundant water create diverse freshwater systems active year-round, many posing flood hazards (Deo et al. 2021; Goldsmith and McCue 2016;

Fig. 4.27 Brochure prepared by the Samoan Tourism Authority showcasing Samoa as an adventure tourism destination

Williams et al. 2021; Wu et al. 2021). The volcanic landscape with basaltic lava flows forms stepwise, deeply incised terrains with various waterfalls especially in high intensity repeated water events (Lamb et al. 2014; Zhang et al. 2020). Recognizing this, the Samoan Tourism Authority promoted sites like Afu Aou Waterfall for tourists to explore and enjoy as recreational spots (Fig. 4.28). Additionally, tube-fed lava flow systems interacting with the high-energy ocean create unique cliff-dominated shorelines with blowholes (Marie 2003), such as those at Alofa'aga Blowholes in southern Savai'i (Fig. 4.29). Blowholes are considered as geoheritage in some cases beside their attractive features to act as tourism visitation sites (Flor-Blanco et al. 2023).

The Samoa Tourism Authority highlights the volcanic geoheritage of the islands, promoting the youngest eruption site (1905–1911) at Mt. Matavanu in Savai'i. Their explanatory board includes geological and biodiversity information (Fig. 4.30).

The Matavanu eruption site shows how local communities or individuals can manage popular locations. The "Crater Man" (Fig. 4.31) guided visitors and created signs around the crater, turning it into a local attraction. The 1905–1911 Matavanu eruption drew global attention as Samoa's youngest volcanic event with lava flows like Hawaii's. This significance earned it a spot on the First 100 IUGS Geological Heritage Sites list (Fig. 4.32). Despite its potential as a geotourism destination, it remains underutilized. A geological summary is available on the IUGS website (https://iugs-geoheritage.org/geoheritage_sites/1905-1911-matavanu-volcanic-eruption/. Accessed 17 December 2024) (Fig. 4.33).

The Matavanu 1905–1911 eruption created a long lava flow field that is easily accessible from Savai'i's main roads. The partially demolished village of Saleaula, recognized by the Samoa Tourism Authority, has become a popular tourist destination (Fig. 4.34). An explanation board and a simple information center, maintained by the local community, provide visitors with detailed insights (Fig. 4.35). Signposts, usually created by locals due to limited central funding, guide visitors Figs. 4.36 and 4.37). Community workshops and joint exercises with geoheritage experts recently promoted these efforts as part of a potential island-wide geopark movement. As an outcome, the Samoa Geopark Program (Fepuleai et al. 2021; Németh et al. 2017) enjoyed some small central support and certainly made the concept visible for the broader Samoan authorities. Although progress is slow, the direction is clear. It is only a matter of time before

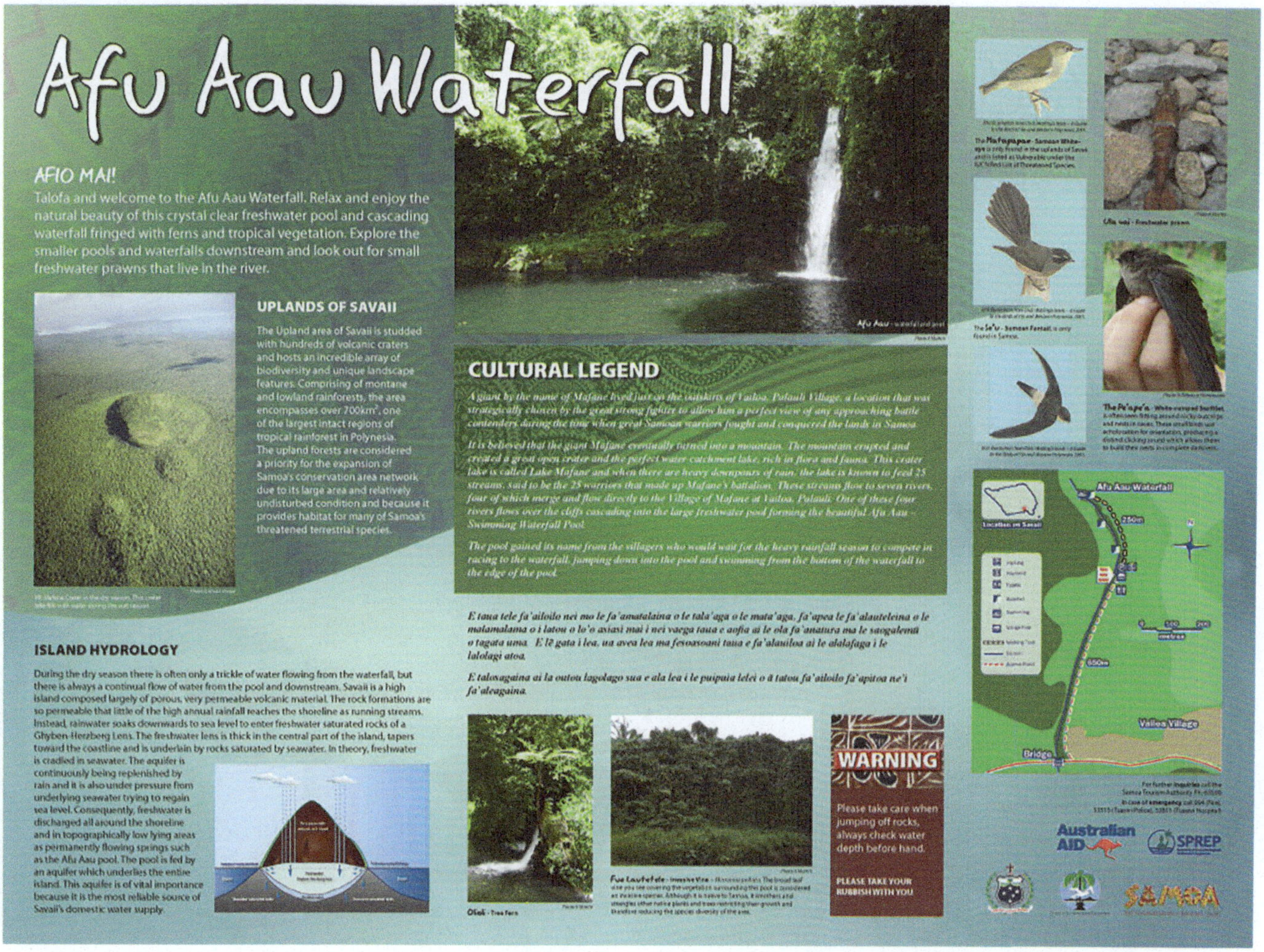

Fig. 4.28 Waterfalls are important landscape elements of Samoa geodiversity, and they provide unique ecosystems beside their geological features. Various aid organizations recently created a set of key sites with simple and scientifically thorough explanatory boards that can be accessed and freely used and redistributed from the website of the Samoan Tourism Authority. (13°44′49″S, 172°18′45″W)

this movement becomes part of Samoa's broader geoheritage, geotourism, and geoconservation strategies.

The Matavanu eruption site presents a significant opportunity for geoheritage within Samoa's tourism sector. It can serve as an additional, abiotic platform to promote and include, but with a properly developed inventory, it can offer a range of alternative geosites for visitors. Although funding is often a challenge in small island nations, relatively low-cost solutions are available. Samoa boasts an extensive and well-maintained mobile network system with internet access from nearly every corner of the islands. By utilizing electronic resources such as QR codes or virtual reality to provide information about geological and geomorphological features, the Samoan Tourism Authority could effectively manage such an inventory. Furthermore, Samoa has a local university, the National University of Samoa (https://nus.edu.ws/. Accessed 17 December 2024), which conducts extensive research on subjects relevant to geocultural and environmental science. Research results and the development of the inventory could be guided with quality assurance from this institution.

The Matavanu eruption site offers a prime opportunity for geotourism development. In south Upolu, O Le Pupu-Pu'e National Park's Coastal Walk is another ideal location for geotourism with minimal investment (Fig. 4.38). Along the south Upolu coastline, numerous potential geosites can be catalogued to support local geotourism ventures. Popular tourist spots like Papase'ea. Sliding Rocks owe their fame to geological features (Fig. 4.39). Highlighting the abiotic nature's role in environmental understanding could further integrate geotourism into Samoa's tourism strategy. The To Sua Giant Swimming Hole, formed by an exposed lava tube, exemplifies how water and volcanism contribute to Samoan tourism (Fig. 4.40).

Samoa has many cultural sites significant to its history. The Samoan Tourism Authority acknowledges Robert Louis

Fig. 4.29 Alofa'aga Blowholes (13°48′05″S, 172°31′17″W) in the southern coast of Savai'i are truly one of the most spectacular sites to see the interaction of wave actions and the longitudinal lava tube network exists within the coastal basaltic lava flows. The various aid agencies provided information laid out in an explanatory board that can be accessed freely from the Samoan Tourism Authority web resources

Stevenson's former homestead as a key destination showcasing early colonial life (Fig. 4.41). This popular site could be expanded with geoheritage-based geotourism, given its location on Mount Vaea, a volcanic scoria cone near Apia. Using existing cultural attractions alongside geological features can enhance tourism opportunities. For instance, the Great Wall of China integrates geosites along its base, drawing more visitors. Proper management and inventory of geosites could significantly boost Samoa's tourism by highlighting its geological heritage.

Fig. 4.30 The volcanic geoheritage of Samoa is significant and the youngest eruption on the islands created a unique landscape in NE Savai'i. The newly created explanatory board accessible freely from the Samoan Tourism Authority can provide a valuable information data set for potential visitors. (13°32′17″S, 172°23′42″W)

Fig. 4.31 The "Crater Man" of the Matavanu eruption site provided services to many visitors in the past decades and created a contemporary heritage venue along the way to the source of the historic eruption site

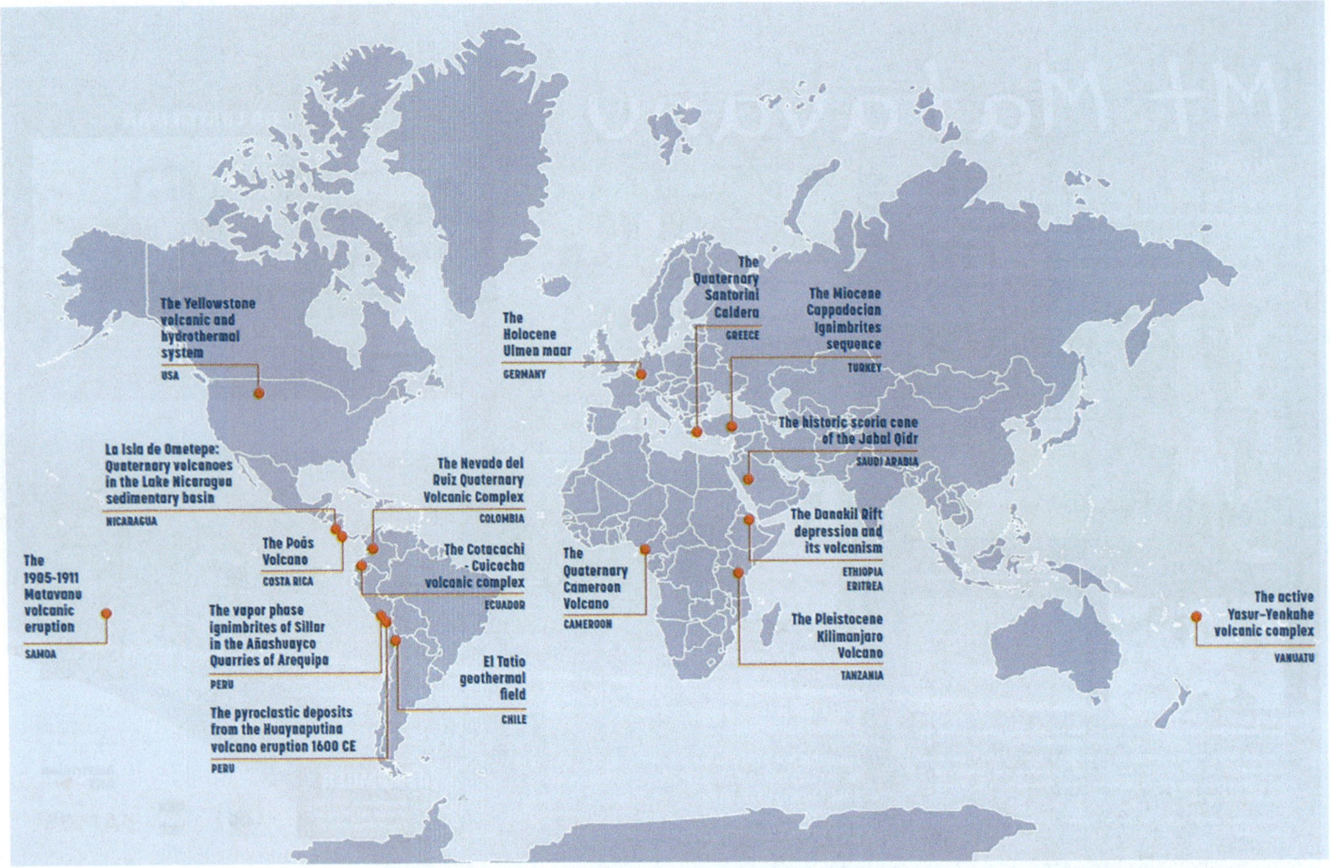

Fig. 4.32 The Matavanu eruption site made it to the list of the First 100 IUGS Geological Heritage sites that is a very prestigious list of significant geological locations on Earth

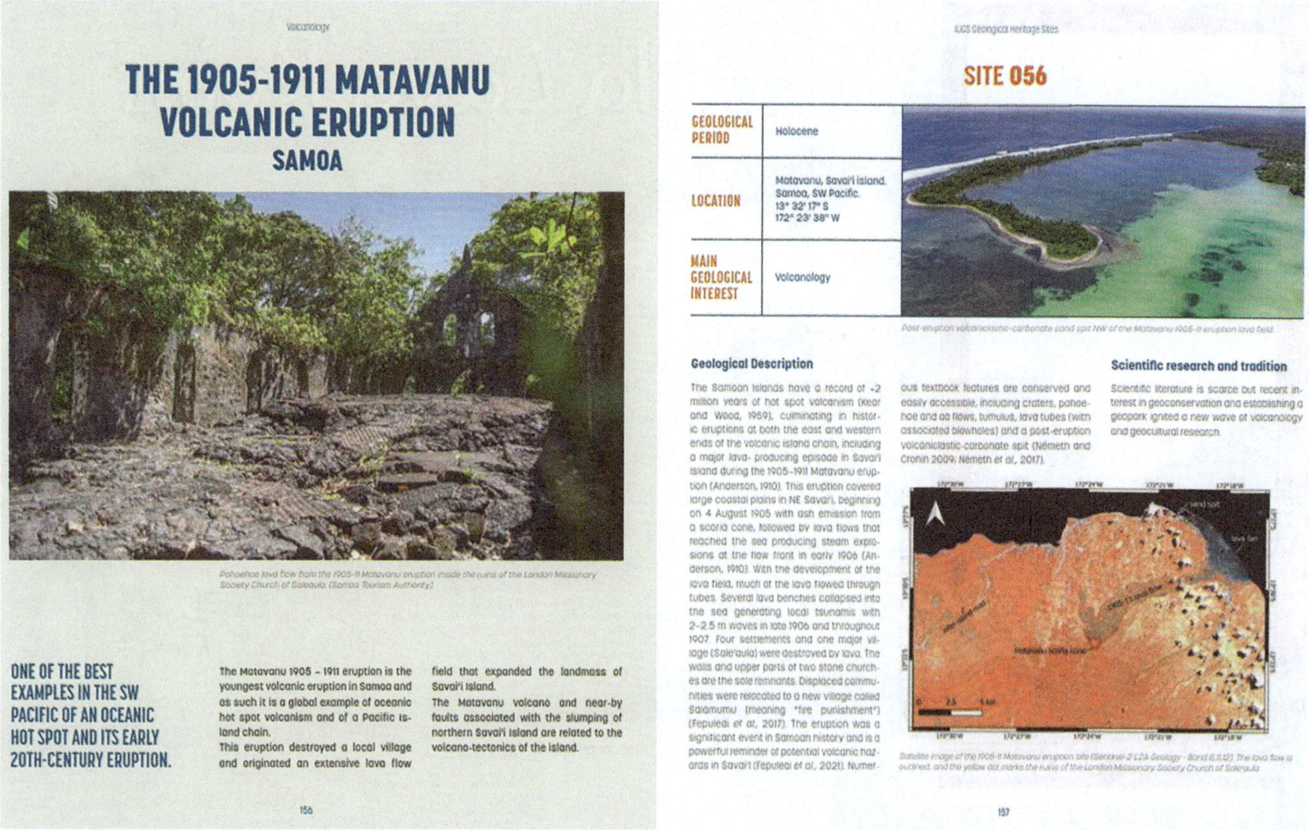
Volcanology

THE 1905-1911 MATAVANU VOLCANIC ERUPTION

SAMOA

Pahoehoe lava flow from the 1905-11 Matavanu eruption inside the ruins of the London Missionary Society Church of Saleaula. (Samoa Tourism Authority)

ONE OF THE BEST EXAMPLES IN THE SW PACIFIC OF AN OCEANIC HOT SPOT AND ITS EARLY 20TH-CENTURY ERUPTION.

The Matavanu 1905 – 1911 eruption is the youngest volcanic eruption in Samoa and as such it is a global example of oceanic hot spot volcanism and of a Pacific island chain.
This eruption destroyed a local village and originated an extensive lava flow field that expanded the landmass of Savai'i Island.
The Matavanu volcano and near-by faults associated with the slumping of northern Savai'i Island are related to the volcano-tectonics of the island.

156

IUGS Geological Heritage Sites

SITE 056

GEOLOGICAL PERIOD	Holocene
LOCATION	Matavanu, Savai'i Island. Samoa, SW Pacific. 13° 32' 17" S 172° 23' 38" W
MAIN GEOLOGICAL INTEREST	Volcanology

Post-eruption volcaniclastic-carbonate sand spit NW of the Matavanu 1905-11 eruption lava field.

Geological Description

The Samoan Islands have a record of ~2 million years of hot spot volcanism (Kear and Wood, 1959), culminating in historic eruptions at both the east and western ends of the volcanic island chain, including a major lava- producing episode in Savai'i Island during the 1905-1911 Matavanu eruption (Anderson, 1910). This eruption covered large coastal plains in NE Savai'i, beginning on 4 August 1905 with ash emission from a scoria cone, followed by lava flows that reached the sea producing steam explosions at the flow front in early 1906 (Anderson, 1910). With the development of the lava field, much of the lava flowed through tubes. Several lava benches collapsed into the sea generating local tsunamis with 2-2.5 m waves in late 1906 and throughout 1907. Four settlements and one major village (Sale'aula) were destroyed by lava. The walls and upper parts of two stone churches are the sole remnants. Displaced communities were relocated to a new village called Salamumu (meaning "fire punishment") (Fepuleai et al., 2017). The eruption was a significant event in Samoan history and is a powerful reminder of potential volcanic hazards in Savai'i (Fepuleai et al., 2021). Numerous textbook features are conserved and easily accessible, including craters, pahoehoe and aa flows, tumulus, lava tubes (with associated blowholes) and a post-eruption volcaniclastic-carbonate spit (Németh and Cronin 2009; Németh et al., 2017).

Scientific research and tradition

Scientific literature is scarce but recent interest in geoconservation and establishing a geopark ignited a new wave of volcanology and geocultural research.

Satellite image of the 1905-11 Matavanu eruption site (Sentinel-2 L2A Geology - Band 8,11,12). The lava flow is outlined, and the yellow dot marks the ruins of the London Missionary Society Church of Saleaula.

157

Fig. 4.33 The summary of the Matavanu 1905–1911 Matavanu volcanic eruption as one of the First 100 IUGS Geological Heritage Sites

Fig. 4.34 The distal sections of the 1905–1911 Matavanu eruption site form the Saleaula Lava Ruins (13°27′7.70″S, 172°19′52.81″W) as a well accessible pristine pahoehoe lava flow field providing great insight to the volcanic eruption mechanism. A well-maintained trail facilitate visitation to the ruins and lava fields

Fig. 4.35 Small information center provides key advice for the walking tours on the lava field. The family operated venture however receives very minimal support from elsewhere (13°27′7.70″S, 172°19′52.81″W)

Fig. 4.36 The explanatory board of the Saleaula Lava Ruin exhibited in the visitor center (13°27′7.70″S, 172°19′52.81″W)

Fig. 4.37 Signposts made by local communities to guide visitors to the Saleaula Lava Ruins (13°27′7.70″S, 172°19′52.81″W)

Fig. 4.38 Information board in the entry of the O Le Pupu-Pu'e National Park, Coastal Walk (14° 2′27.97″S, 171°45′2.05″W)

Fig. 4.39 Information board of the Papase'ea. Sliding Rocks near Apia, Upolu (13°52′10″S, 171°48′17″W)

Fig. 4.40 Information board of the To Sua Giant Swimming Hole (Upolu) because of some skyline opening on a large lava tube network subsequently occupied by the ocean (14°02′38″S, 171°33′44″W)

Fig. 4.41 Historic site of the Robert Louis Stevenson Memorial on Mt. Vaea Scenic Reserve within the township of Apia, Upolu. The site is on an older scoria cone that dominates the landscape of Apia (13°51′59″S, 171°45′56″W)

4.3 Conservation Programs

In Samoa, there are a strong emphasis on conservation programs that are also aligned with many global attempts. The programs however are almost exclusively aligned with various biodiversity conservation projects or environmental science or global and planetary changes relevant research and outreach. Most of the programs are oversighted by the Ministry of Natural Resources and Environment (MNRE) (https://www.mnre.gov.ws/. Accessed 15 December 2024). Official running project includes Enhancing Resilience to Climate Change (), Marine Spatial Project (), Impress Project (), and Works as Customary Land Advisory Commission () as a major conversation-related works that enjoys government supports and aligned with medium to long-term government goals. The MNRE also involved in more target specific programs such as fighting against alien species (GEF-7 Samoa Invasive Alien Species (IAS) Project). The Marine Spatial Projects focuses on the preservation and sustainable management of the extensive marine environment Samoa occupies (https://www.mnre.gov.ws/?page_id=945. Accessed 12 December 2024). In this project while it is more biodiversity focused, there are components on understanding the complex coastal shallow marine environment and has clearly elements of geoconservation opportunity, despite it has not mentioned anywhere in the information packages.

As Samoa facing a complex geohazard from volcanic hazard to earthquakes and associated tsunamis, or from extreme weather, sea level-rise related coastal processes and flooding, MNRE has a complex role to conduct research, observe and mitigate such geohazards (https://www.mnre.gov.ws/?page_id=250. Accessed 15 December 2024). This program has its own section to provide link to the society and manage information channels to local communities. Within these programs, there is no explicit mention on geoconservation or the potential to use geoheritage and geodiversity to geohazard resilience programs as suggested elsewhere, but this might change quickly once the concept infiltrate to the organization and a natural need emerges.

MNRE is also the national organization to oversight environmental and conservation issues (https://www.mnre.gov.ws/?page_id=254. Accessed 15 December 2024). MNRE is a long-lasting organization within conservation responsibilities. The Division of Environment & Conservation (DEC) responsible to deal and oversight issues with different emerging problems that has threatened Samoa's biodiversity from land to sea. Within this Division major conservation areas are handled by the National Parks, Terrestrial Conservation and the Marine Conservation offices. As it stated, and also visible from the actions of this office the main responsibility of it to implement activities and programs that tries to address the diversity of threats impacting Samoa's biodiversity (https://www.mnre.gov.ws/?page_id=254. Accessed 15 December 2024). While there were and there are activities that clearly could be listed as geodiversity relevant programs, the word is not mentioned in anywhere. Unfortunately, this is not Samoa-specific program, as geodiversity is still an emerging concept for most of the countries, and only few countries adopted full it to express its importance in government level conservation strategies. DEC's overall objective is to "implement sustainable development and management of biological resources and the environment including waste management".

Samoa biodiversity profile is very impressive and it is clear that lots of effort was taken to maintain the preservation and conservation coals (https://www.cbd.int/countries/profile/default.shtml?country=ws#status. Accessed 15 December 2024). According to the Convention on Biological Diversity (https://www.cbd.int/convention. Accessed 15 December 2024), the main challenges Samoa faces summed up as "*The main threats to biodiversity identified in the NBSAP include: forest clearance, population growth, overexploitation of natural resources, non-sustainable development, natural disasters and the spread of introduced animal and plant pests. More specific threats can also be mentioned, such as climate change, coastal pollution, logging, agricultural clearing, forest fires and human settlement. The underlying drivers of change and causes of threats include economic, demographic, institutional, technological, cultural, social and political factors*". It is evident that within these challenges, most of them has strong geodiversity elements, and certainly would work better to treat them as a complex geo-biodiversity issue as solutions might be more sustainable. Climate change and its broader aspects as global and planetary changes has clear aspects on coastal morphology changes, lagoon and reef reduction, landslides, flooding, mass movements or even volcanic or seismic hazards.

Samoa national park and reserve program and management is also entirely under the MNRE management. Historically, the largest National Park (by surface area), O Le Pupu-Pu'e National Park in southern Upolu (This Samoan name means "The Cliffs-Mountain") (Fig. 4.42), has been

Fig. 4.42 The entry to the O Le Pupu Pu1e National Park that is also partially overlap with a RAMSAR region forming a significant nature conservation area of southern Upolu (14°00′56″S, 171°43′19″W)

the first initiative of a national park program to start in Samoa (Trotman 1979). The goals of this national park summarized as "(1) Conserve a large representative area of the landforms, vegetation and wildlife in a largely unmodified state. (2) Conserve water supplies for current and future use. (3) Communicate to the public the concepts and values of the natural history of the country and their natural heritage, providing an outdoor classroom. (4) Provide tourist attractions containing features of outstanding interest and representation of the natural environment" (Trotman 1979). Among these four goals two are clearly geoconservation, geoheritage and geodiversity relevant goals. A coastal walkway with some explanatory boards clearly demonstrates that geology and geomorphology still in the sight of the management, however it is little bit hidden and not promoted in a level, such an intact landscape could be.

As water is an increasingly threated reserve even in a tropical island MNRE put lots of effort to education on the importance to preserve water resources (Fig. 4.43). Alongside these programs various international aid agencies also involved in small projects to highlight the importance to conserve wetlands and follow sustainable developments by using natural resources for wastewater managements. Nice example for this the establishment of garden toilets along major tourism attractions that are also become a tourist visitation spots beside their sustainable functionality (Figs. 4.44 and 4.45).

Samoa has a well-established ocean conservation management strategy labelled as the Samoa Ocean Strategy that has been finalized, signed and launched in 2020 (https://www.samoaocean.org/. Accessed 15 December 2024). As outlined in their mission statement.

> The Samoa Ocean Strategy, or SOS, is the national policy framework that seeks to sustainably manage Samoa's vast ocean and marine resources for the well-being of all Samoans now, and into the future.

This is a very broad and overarching program centered around the Pacific Ocean, its resources, resource management and conservation. The SOS identifies and aims to safeguard the cultural, ecological and socioeconomic values that Samoans derive from their ocean including (1) fisheries, (2) tourism, (3) offshore waters, (4) maritime safety and security, (5) biodiversity and ecosystems, (6) food security, (7) ocean knowledge and traditions. Geoconservation, geodiversity or geoheritage is not mentioned in the strategy, but at least within tourism, ecosystems and ocean knowledge and tradition's theme host significant aspects of "geo" themes where synergy would generate sustainable programs. This is especially valid for expanding ecosystem protections to geosystems incorporating the ecosystem and geosystem services concepts (Arias-Díaz et al. 2023; Attard 2019; Bobylev et al. 2022; Chen et al. 2024; Fox et al. 2020; Frisk et al. 2022; Gray 2011, 2012; Miklós et al. 2020; Németh et al. 2021a, b).

Fig. 4.43 Common boards keeping attention on the importance to maintain environmental conditions, especially water resources intact

Fig. 4.44 Garden toilet in the start of the walking track to the La'auolola lava caves (13°32′18″S, 172°31′45″W) within the Aopo Forest Reserve and Cave and Forest Toilet reserve

The SOS provides bold and comprehensive integrated ocean management solutions that will advance ocean stewardship and ensure the cultural and the economic values that Samoans derive from their 120,000 km^2 ocean are preserved for generations to come.

The SOS successfully established protected areas such as the Aleipata Districts Marine Protected Area (MPA) or four mangrove reserves established with management plans finalized, endured and officially launched across the islands. The continuation of these projects to expand it to other vulnerable areas are on the agenda, however, it is still now explicit declaration to have geoconservation as a vital part of such project. This is an area where development is expected, as the 2009 tsunami showed for instance how important coastal mangrove forests are for reducing the wave impact on such geodisasters. Also, mangrove forests are complex geosystems as well as they play important role in the terrestrial sediment capture and mud plain formation, forming a unique sedimentary environment, that has high geoheritage value (Huber 1994; Kathiresan 2021; Risanti and Marfai 2020; Rogers 2021; Saintilan and Rogers 2013; Sukojo and Annisa 2022; Zhang et al. 2013).

Beside marine biodiversity projects various terrestrial biodiversity conservation projects operated in Samoa under the NRME. However, these projects are less apparent and visible than those in the marine environment.

One of the most relevant currently operating conservation venture that has strong geoconservation connotation is the Pacific Islands Ridge-to-Reef program (https://www.pacific-r2r.org/. Accessed 15 December 2024). The goal of this program is outlined as below.

> To maintain and enhance Pacific Island countries' ecosystem goods and services (provisioning, regulating, supporting and cultural) through integrated approaches to land, water, forest, biodiversity and coastal resource management that contribute to poverty reduction, sustainable livelihoods and climate resilience.
>
> The Pacific Ridge to Reef programme is a Global Environment Facility (GEF) multi-focal area programme guiding coordinated investment of GEF grant funding across its focal areas of biodiversity conservation, land degradation, climate change adaptation and mitigation, sustainable land management, sustainable forest management, and international waters in Pacific SIDS. It is a multi-agency initiative involving the United Nations Development Programme (UNDP), the United Nations Food and Agriculture Organization (FAO), and the United

Fig. 4.45 Garden toilets as part of the World Wetland Day projects in Upolu near the walking track start of the O Le Pupu-Pu'e National Park (14°00′56″S, 171°43′19″W) toward the Togitogiga Waterfall

> Nations Environment Programme (UNEP) as GEF implementing agencies.
>
> Executed regionally by the Pacific Community through the GEF Pacific Ridge to Reef International Waters project (GEF Pacific R2R IW), the operations of the R2R programme is supported in areas of science-based planning, human capital development, policy and strategic planning, results-based management, and knowledge sharing. Implemented through the GEF Pacific R2R IW project the R2R IW national pilot projects are designed to strengthen R2R integration by establishing synergies between sector agencies and the GEF National R2R STAR Projects, governments and communities, civil society and the private sector.
>
> The Programme Coordinating Unit (PCU), hosted by the Pacific Community's Geoscience Division based in Fiji, is tasked with the provision of technical, operational, reporting and monitoring support as requested by the participating PacSIDS.

The outline of the R2R is very promising and there is plenty of good connecting places to apply not only ecosystem but also geosystem services to the functionality of the program. It is also noticeable however, despite the program operates within diverse geoenvironment, geoconservation or geodiversity, neither geoheritage appeared or spelled out in the documents. It is very evident that this program could possibly be one of the best avenues to establish a geoconservation perspective not only in Samoa but across the Pacific. "*The R2R program promotes the conservation and sustainable use of biodiversity and the maintenance of ecosystem goods and services through the improved management of existing and new protected areas, sector reforms to conserve and sustainable use biological diversity, and the incorporation of biodiversity conservation and sustainable use into planning frameworks.*" As in many cases, R2R is also heavily biased toward biodiversity conservation as the key for a broader socio-economically important conservation strategy. This is rather interesting as the Biodiversity Management within R2R clearly includes areas commonly viewed as subject areas of geoconservation. This is a particularly important subject to note as climate change or anything associated with it and impacting land use changes, geohazards and environmental protection based on the preservation of the abiotic nature, hence geoconservation.

4.4 Place of Geoconservation

Samoa has advanced conservation strategies, primarily focusing on biodiversity. However, geoconservation could also benefit areas like coastal habitat preservation, land-use change monitoring, soil erosion, extreme flooding events, and volcanic eruptions. Expanding existing programs to include geoconservation might be feasible if local experts understand its importance.

Geoconservation involves identifying, protecting, and managing sites of geological or geomorphological value (Bridgland 2013; Brilha 2016; Brilha and Pereira 2020; Brocx and Semeniuk 2007; Gordon and Barron 2012; Henriques et al. 2011; Larwood et al. 2013; Monge-Ganuzas et al. 2024; Moura et al. 2017; Prosser 2012; Reynard and Brilha 2017). Government agencies and geological societies across the Globe practice this by analyzing sites and implementing management infrastructure (Allan Donaldson 2009; Asrat et al. 2012; Carcavilla et al. 2009; Comfort and Eberhard 2011; Ellis 2011; Gonera 2010; Liu et al. 2010; Parkes 2008; Prosser 2011; Reynard et al. 2021; Seijmonsbergen et al. 2010; Worton 2008). The principles aim to protect these sites and evaluate their significance to the geological community. Geodiversity conservation often occurs alongside biodiversity conservation. Protected "geosites" contribute to Earth heritage and illustrate geological processes, promoting public understanding, education, and research.

The principles of geoconservation involve creating an inventory of geoheritage sites, assessing their value, managing their conservation, monitoring the geoconserved sites, and promoting the sites through interpretation. Field specialists add geosites to the inventory by conducting surveys, mapping, and fieldwork. The site documentation involves collaboration with non-specialist landowners, rangers, and protected area managers. In Samoa, it is logical for the Ministry of Natural Resources and Environment (MNRE) to act as a custodian of the geosite inventory, given its government status and strong connections with communities and other stakeholders.

The primary objective of geoconservation is to protect areas designated for conservation by evaluating the geological structures and composition of the sites. As geoconservation operates under similar legislative frameworks as other conservation entities, MNRE in Samoa is well-suited to serve as custodian. The value of a site is primarily determined by its scientific significance, along with its cultural and aesthetic importance, which enhances its appeal for geotourism and biodiversity conservation. Geomorphological sites to be conserved may include static features like relict volcanic landforms or active features such as coastal (coral reef fringed lagoons) and river landforms.

A comprehensive assessment of the site's management can provide insights into protecting and maintaining the site, ideally incorporating biodiversity and ecosystem assets. Management plans should address the site's suitability for geotourism, including an outline of geotourism management strategies (Erikstad 2013; Fassoulas et al. 2012; Fuertes-Gutiérrez and Fernández-Martínez 2010; Kubalíková and Balková 2023; Lima et al. 2013; Migoń and Pijet-Migoń 2024; Raposo et al. 2018; Santos-González and Marcos-Reguero 2019; Zuo and Fang 2007). These plans must include information about potential geohazards that could affect visitors or damage the geosites (Avagyan et al. 2023; Herrera-Franco et al. 2023; Németh et al. 2021a, b; Novković et al. 2022; Sajinkumar 2013). Furthermore, management plans should be adaptive, particularly for island nations like Samoa where frequent earthquakes, floods, and coastal erosion from cyclones can impact geosites under geoconservation programs.

Geosites are zoned to facilitate proper management of protected areas. Core areas within the geoheritage site are prioritized for protection and require significant infrastructure, monitoring, or restoration plans. Buffer zones surrounding the core areas support these efforts and allow for a wider range of activities. To maintain the integrity of the geoconservation area, site condition monitoring is conducted to ensure compliance with conservation objectives. Regularly scheduled assessments help ensure the effective management and conservation of the sites. This type of management plan can easily be adapted to general conservation plans, accommodating the specific features of geosites.

Geoconservation is essential in developing geoparks, from National Geoparks to UNESCO Global Geoparks (Alexandrowicz 2007; Alexandrowicz and Alexandrowicz 2004; Canesin et al. 2020; Chakraborty et al. 2014; Erikstad 2008; Fung and Jim 2015; Gordon 2019; Henriques et al. 2012; Lima et al. 2014; Ólafsdóttir and Dowling 2014; Piranha et al. 2011; Reynard and Brilha 2017; Turner 2013; Vilas Boas et al. 2013). Identifying and protecting key geological heritage sites are vital for any geopark's strategies. However, sustainable economic development also considers the demands of scientific, commercial, and tourism uses of these sites. Historically, competing interests have sometimes led to inadequate protection for significant geological sites.

Geoparks are crucial in highlighting the significance of geodiversity and in educating the public and policymakers about its comprehensive protection (Carrión-Mero et al. 2024a, b; El hamidy et al. 2024; Grigorescu 2020; Mehdioui et al. 2022; Naimi and Cherif 2021; Quesada-Valverde and Quesada-Román 2023; Rosado-González et al. 2020; Simbaña-Tasiguano et al. 2024). They spearhead the conservation and sustainable promotion of geoheritage by identifying and safeguarding geosites. Through long-term sustainable management and geoconservation initiatives, these sites can be preserved for current and future generations, ensuring their availability for educational purposes, including community geohazard resilience development programs.

In Samoa, the Samoa Geopark Program was launched to assess the feasibility of establishing the first geopark in the Southwest Pacific islands (Fepuleai et al. 2021). The program concentrated on Savai'i's unaltered natural landscapes from the coast to the peaks and identified potential geosites. It also evaluated community responses to the concept and proposed several strategic plans. Currently, the program is on hold due to disruptions caused by COVID-19 and insufficient financial and intellectual support.

Despite these challenges, geoparks play a crucial role in geoconservation, making it essential to revive the work already started in Samoa. While there is a common perception that geoparks must be global and international, numerous examples highlight the importance of local or regional (national) geopark networks. For those geoparks aspiring to higher recognition and inclusion in the UNESCO Global Geopark Network (https://www.unesco.org/en/iggp/geoparks/about. Accessed 19 December 2024), a thorough evaluation and submission of dossiers to UNESCO are required. Each UNESCO Global Geopark, as part of their submission documentation to UNESCO, must demonstrate the significance of geosites within their territory. This involves having geological heritage of international importance, which means at least one designated site of international significance within the territory, along with multiple nationally and regionally important sites that collectively support the themes of the geopark submission.

Initial studies have shown that Samoa has numerous geosites of international significance, including one listed on the IUGS First 100 Geological Heritage Sites (https://iugs-geoheritage.org/. Accessed 18 December 2024). Thus, Samoa could initially develop a national geopark, which may later transition into a UNESCO Global Geopark.

A geopark is an area that integrates geological heritage with protection, education, and sustainable development. It encompasses not only sites of geological importance, but also ecological, archaeological, historical, and cultural sites connected to the landscape and geology. In Samoa, natural and cultural history are intricately linked, influencing daily life.

Geoparks aim to conserve significant geological features such as basaltic lava flows, peridotite lherzolites, and landforms that inform various geoscientific disciplines, including geomorphology, hydrology, petrology, and volcanology. They explore conservation methods and promote sustainability through education, ensuring effective protection and maintenance in consultation with statutory bodies. Geoparks must inventory natural resources and adopt a holistic approach, combining geo- and biodiversity management strategies. Samoa exemplifies this approach and could benefit from incorporating it into its geopark strategy.

The preservation of geological heritage in on-site museums or interpretation centers can serve as an attraction for tourists and increase visitors to the geopark. These facilities offer educational programmes for students and teachers at all educational levels. Additionally, there are opportunities to produce and sell thematic products related to the geological features of the geopark. These products may include (1) replicas of rocks, fossils, and minerals from sustainable sources controlled by the geoconservation management entity (e.g., scoria, pahoehoe lava, volcanic bomb), (2) symbolic products for local cafes and restaurants (e.g., branding using Samoan geoheritage), (3) mascots (e.g., specific volcanic forms or shells), (4) handicrafts (e.g., artforms featuring lava flow or printing volcano shapes on traditional Samoan cloths like lava lavas), (5) artistic prints or photographs (e.g., images framed in volcanic rocks, or historical images from the 1905–1911 Matavanu eruption), or (6) geo-foods (https://geofood.no/. Accessed 18 December 2024) associated with geology (e.g., marketing umu cooked using heated basaltic rocks, resembling geology-themed pastries). Geological heritage can also inspire cultural activities (theatre, music, dance) and the creation of educational materials. The connection between volcanic geoheritage and Samoan cultural traditions is clear but not often explicitly shown (Fepuleai et al. 2017). During dance shows, these connections could be highlighted to help visitors understand the link between geoheritage and daily life. A comprehensive geoconservation strategy creates sustainable markets, improves job prospects, and generates economic benefits for local communities, contributing to UNESCO's sustainable development goals.

Geoconservation action strategies for geoparks typically focus on:

- Site conservation: Maintain and update geosite registers, provide evidence for marine, volcanic, and coastal conservation in Samoa, advise or manage site conservation, and support sites under threat.
- Developing geoconservation agendas: Engage with new conservation priorities like ecosystem services, marine conservation, and understand links between geodiversity and biodiversity for a holistic approach to nature conservation and environmental management.
- Interpretation, outreach, and policy influence: Collaborate with local geoconservation groups and traditional Samoan communities, respond to government consultations on planning and environmental issues, and support international geoconservation developments.

Geoconservation in a geopark involves sharing best practices through partnerships, field meetings, conferences, outreach, and publications. Some organizations now have dedicated geoconservation officers.

A geoconservation strategy should include:

- Auditing sites and collections
- Engaging with national bodies

- Proposing site conservation
- Managing access and visitor expectations
- Identifying interpretation opportunities

Geoconservation allows local groups to adopt and maintain geosites, such as:

- Clearing vegetation and loose material from rock faces
- Maintaining paths
- Contributing cultural knowledge to interpretation projects
- Acting as wardens or rangers

These actions are well-aligned with Samoan cultural practices and can be effectively integrated into any geopark on the island.

A critical aspect of geoconservation involves managing visitor numbers and movements at geosites. Car parking should be strategically located to avoid detracting from the geosite, and alternative travel methods should be encouraged when feasible. Pathways should respect the local landscape and be constructed using local materials to blend in with the geosite; curving paths that follow the natural topography may be more suitable than straight ones. Interpretation panels, which serve as focal points, require careful placement to prevent rapid erosion due to footfall. In some cases, it may be necessary to limit monthly or annual visitor numbers to prevent site damage or tension within the local community. Visitor management is essential for geosite conservation, depending on the site's nature. The geopark management authority must also anticipate subsequent maintenance to mitigate visitor impact and ensure resources are available to prevent deterioration of the site and its surroundings.

In Samoa, visitation to natural geosites is currently very low, making visitor control unnecessary. However, this situation may change once a geopark is established. Conversely, the low visitation rate necessitates regular geotrail maintenance, as tropical climates lead to frequent revegetation, landslides, or flooding. This issue is significant for Upolu's Lanoto crater lake track maintenance, where overgrowth within a few months can pose substantial hazards for visitors navigating the tropical rainforest.

Samoa showcases geoheritage as a cornerstone of tourism, supported by community pride in keeping villages clean and appealing. Sites like Pe'a Cave in northern Savai'i are part of tourism programs managed by local communities (Fig. 4.46). The cave features an informative board explaining its geological, biological, and cultural significance

Fig. 4.46 Geological explanation board in the entry of the Pe'a Cave that is a lava tube network in northern Savai'i. The explanatory board is in a private land and the cave maintained and managed by local family (13°30′23″S, 172°28′40″W)

Fig. 4.47 Well maintained entry with typical Samoan style flower arrangements to the Pe'a Cave (13°30′23″S, 172°28′40″W)

(Fig. 4.46). The well-maintained entrance, surrounded by local flowering plants, offers a relaxing tropical atmosphere (Fig. 4.47).

Moso's Footprint, signposted from the main road to the NW corner of Savai'i, is a notable site (Fig. 4.48). It showcases a lava surface texture common on pahoehoe lava flows, formed by cooling and solidification that creates slightly domed areas prone to fracturing (Fig. 4.49). Continuous magma supply allowed gases to escape through the solidifying crust, forming a chilled, vesicular flow top. Post-eruption, weathering can create patterns resembling footprints. Local legends attribute these formations to Moso's steps, embedding the geosite in Samoan culture. Maintained by the community with informative boards, it attracts many visitors (Fig. 4.50). While Samoa has similar sites, enhancing geological information and conservation efforts could help develop a national network of geosites, potentially leading to the recognition as a UNESCO Global Geopark.

Fig. 4.48 Moso's Footprint in NW Savai'i an important location that has a legend associated with. It importance is highlighted by the fact that its access is well marked on the road (13°30′05″S, 172°46′01″W)

Fig. 4.49 Moso's footprint is a typical lava surface feature that shaped like a large footprint. The location provides good summary on the legend with a basic geological summary (13°30′05″S, 172°46′01″W)

Fig. 4.50 Explanatory board of the Moso's Footprint as an important geocultural location in NW Savai'I (13°30′05″S, 172°46′01″W)

4.5 Web Resources on Conservation in Samoa

The below URLs were accessed on 15th January 2025.

https://star.gsd.spc.int/
https://star.gsd.spc.int/index.php/joomla-pages/abstracts
https://www.nps.gov/articles/nps-geodiversity-atlas-national-park-of-american-samoa-national-park-american-samoa.htm
https://www.conservation.org/samoa
https://www.conservation.org/samoa/blog/samoa-voyaging-society-signed-on-to-manaaki-project
https://www.conservation.org/samoa/blog/samoa-s-marine-spatial-plan-releases-informative-reports-and-national-awareness-campaign
https://www.conservation.org/samoa/blog/samoa-first-in-the-pacific-to-use-seasketch-spatial-tool
https://www.conservation.org/samoa/blog/the-voyage-to-managing-100-of-samoa%E2%80%99s-ocean-domain
https://www.samoatourism.org/Content/SiteResources/PAGE/48/TOUR%20GUIDES%20MANUAL.pdf
https://www.samoatourism.org/articles/49/visitors-manual-a-geological-history
https://www.samoatourism.org/articles/46/attraction-sites
https://www.samoatourism.org/section/67/interpretive-signs
https://www.pacific-r2r.org/resource-library?term_node_tid_depth=16
https://www.pacific-r2r.org/sites/default/files/2020-03/Project_Progress_Samoa.pdf
https://pacificdata.org/data/eu/organization/samoa-data?topic=Environment&license_id=sprep-public-license&member_countries=ws&tags=gis
https://www.samoaocean.org/about
https://www.pacific-r2r.org/
https://www.pacific-r2r.org/resource-library?term_node_tid_depth=16
https://www.pacific-r2r.org/sites/default/files/2020-03/Project_Progress_Samoa.pdf
https://pacificdata.org/data/eu/organization/samoa-data?topic=Environment&license_id=sprep-public-license&member_countries=ws&tags=gis

References

Agarwal S (1997) The resort cycle and seaside tourism: an assessment of its applicability and validity. Tour Manag 18(2):65–73

Agarwal S (2002) Restructuring seaside tourism: the resort lifecyle. Ann Tour Res 29(1):25–55

Ahdar RT (2013) Samoa and the Christian state ideal. Int J Study Christ Church 13(1):59–72

Alexandrowicz Z (2007) Geoconservation in national, European and global aspect (with particular regard to Poland). Biul Państw Inst Geol 425:19–26

Alexandrowicz Z, Alexandrowicz SW (2004) Geoparks - the most valuable landscape parks in Southern Poland. Pol Geol Inst Spl Pap 13:49–56

Ali-Knight J, Ensor J (2017) Salute to the sun: an exploration of UK Yoga tourist profiles. Tour Recreat Res 42(4):484–497

Allan Donaldson J (2009) Geoheritage 2. Examples of geoeducation, geoconservation and geo-rescue projects in ontario. Geoscience 36(3):102–106

Alofaituli B (2023) Religion and development in Samoa: time to draw on the strength of local culture? Dev Policy Rev 41(S2):e12746

Arias-Díaz A, Murcia H, Vallejo-Hincapié F, Németh K (2023) Understanding geodiversity for sustainable development in the Chinchiná River Basin, Caldas. Land 12(11):2053

Asrat A, Demissie M, Mogessie A (2012) Geoheritage conservation in Ethiopia: the case of the simien mountains. Q Geogr 31(1):7–23

Attard M (2019) The sustainability of landforms and landscapes. In: World geomorphological landscapes. Springer, Cham, pp 373–380

Aumua A, Fulu- Aiolupotea SM, Suaalii- Sauni T, Fadgen T (2023) Navigating the early impacts of covid-19: south pacific viewpoints from fiji and samoa. In: Democracy, state capacity and the governance of COVID-19 in Asia-Oceania. Routledge, London, pp 113–130

Avagyan A, Sahakyan L, Meliksetyan K, Hovhannisyan A, Arakelyan D, Galoyan G, Melik-Adamyan H, Grigoryan T, Sahakyan K, Grigoryan E, Avagyan S, Safaryan R (2023) The potential for a geohazard-related geopark in armenia. Geoheritage 15(4):133

Barutçu S, Doğan H, Üngüren E (2011) Tourists' perception and satisfaction of shopping in alanya region: a comparative analysis of different nationalities. Procedia Soc Behav Sci 24:1049–1059

Battilani P (2024) Tourism and the blue economy. Italia Contemp 305:192–212

Berrezueta E, Sánchez-Cortez JL, Aguilar-Aguilar M (2021) Inventory and characterization of geosites in ecuador: a review. Geoheritage 13(4):93

Blumrodt J, Palmer A (2014) On-line destination branding: an investigation into the divergence between brand goals and on-line implementation. J Appl Bus Res 30(6):1597–1606

Bobylev N, Syrbe RU, Wende W (2022) Geosystem services in urban planning. Sustain Cities Soc 85:104041

Bouvet Y, Coccaró JM, Desse RP (2001) The shadow of a metropolis: coastal tourism in the province of Buenos Aires (Argentina). Aust Geogr 128:39–54

Bridgland DR (2013) Geoconservation of quaternary sites and interests. Proc Geol Assoc 124(4):612–624

Brilha J (2016) Inventory and quantitative assessment of geosites and geodiversity sites: a review. Geoheritage 8(2):119–134

Brilha J, Pereira P (2020) Geoconservation in Portugal with emphasis on the geomorphological heritage. In: World geomorphological landscapes. Springer, Cham, pp 307–314

Brocx M, Semeniuk V (2007) Geoheritage and geoconservation - history, definition, scope and scale. J R Soc West Aust 90:53–87

Buckwell A, Ariki MP, Oiire C, Unga C, Fleming C (2024) Twenty five years of world heritage status: show us the benefits! J Environ Manag 358:120849

Bull C, Weed M (1999) Niche markets and small Island tourism: the development of sports tourism in Malta. Manag Leis 4(3):142–155

Cameron E (2024) Objectivity in geosites inventorying and assessment methods. Geoheritage 16(1):30

Camilleri MA (2018) The tourism industry: an overview. In: Tourism, hospitality and event management. Springer, Milan, pp 3–27

Canesin TS, Brilha J, Díaz-Martínez E (2020) Best practices and constraints in geopark management: comparative analysis of two Spanish UNESCO global geoparks. Geoheritage 12(1):14

Carcavilla L, Durán JJ, García-Cortés A, López-Martínez J (2009) Geological heritage and geoconservation in Spain: past, present, and future. Geoheritage 1(2–4):75–91

Carrión-Mero P, Dueñas-Tovar J, Jaya-Montalvo M, Herrera-Franco G, Berrezueta E, Morante-Carballo F (2024a) Assessment of UNESCO global geoparks websites for a public geocommunication. Int J Geoheritage Parks 12(2):223–240

Carrión-Mero P, Sánchez-Zambrano E, Mata-Perelló J, Jaya-Montalvo M, Herrera-Franco G, Berrezueta E, Espinel RL, Baque M, Morante-Carballo F (2024b) Geosites assessment in a volcanic hotspot environment and its impact on geotourism, Santa Cruz-Galapagos Islands, Ecuador. Int J Geoheritage Parks 12(1):147–167

Chakraborty A, Cooper M, Chakraborty S (2014) Geosystems as a framework for geoconservation: the case of Japan's Izu Peninsula Geopark. Geoheritage 7(4):351–363

Chen H, Sloggy MR, Escobedo F, Koskimäki T, Lu T, Meng Z, Rasheed AR, Sánchez JJ, Tan X, Yang W, Yu F (2024) Boundary of ecosystem services: differentiating between ecosystem services and geosystem services is needed. J Environ Manag 362:121285

Cheung LTO (2016) The effect of geopark visitors' travel motivations on their willingness to pay for accredited geo-guided tours. Geoheritage 8(3):201–209

Cleal CJ, Thomas BA, Bevins RE, Wimbledon WA (1999) GEOSITES - an international geoconservation initiative. Geol Today 15(2):64–68

Comfort M, Eberhard R (2011) The Tasmanian geoconservation database: a tool for promoting the conservation and sustainable management of geodiversity. In: Proceedings of the Linnean Society of New South Wales, pp 27–36

Connell J (2009) Birdwatching, twitching and tourism: towards an Australian perspective. Aust Geogr 40(2):203–217

Connell J (2021) COVID-19 and tourism in Pacific SIDS: lessons from Fiji, Vanuatu and Samoa? Round Table 110(1):149–158

Constantin CP, Papuc-Damaşcan V, Blumer A, Albu RG, Suciu T, Candrea AN, Ispas A (2021) Profiling visitors to romanian ecotourism destinations. Sustain For 13(5):2958

Côté JE (1997) A social history of youth in Samoa: religion, capitalism, and cultural disenfranchisement. Int J Comp Sociol 38(3–4):217–234

Cvetanović M, Gajić M, Kovjanić A, Krstić F, Sedlak M (2024) Geosite assessment as a contribution to the sustainable development of Babušnica, Serbia. Open Geosci 16(1):644

D'Hauteserre AM (2016) Ecotourism an option in small island destinations? Tour Hosp Res 16(1):72–87

Dé LL, Jackson-Becerra FJ (2021) COVID-19 and transnational remittances in Samoa: maintaining family ties in the Face of Crisis. In: COVID in the Islands: a comparative perspective on the Caribbean and the Pacific. Springer, Cham, pp 279–297

De La Torre GMV, Pérez LM (2017) Religious tourism in different geographical areas in Spain: the tourist profile. Bol AGE 75:29–54

Deo A, Chand SS, Ramsay H, Holbrook NJ, McGree S, Magee A, Bell S, Titimaea M, Haruhiru A, Malsale P, Mulitalo S, Daphne A, Prakash B, Vainikolo V, Koshiba S (2021) Tropical cyclone contribution to extreme rainfall over southwest Pacific Island nations. Clim Dyn 56(11-12):3967–3993

Dim CIP (2024) Creating a geosite inventory for some classic outcrops in Nigeria's Southeastern Province. Geoheritage 16(2):59

Diniz MT, de Araújo IGD (2022) Proposal of a quantitative assessment method for viewpoint geosites. Resources 11(12):115

Do Santos BH, Fonseca Filho RE, Castro PTA (2021) Potential urban geotourist profile of ouro preto (MG, Brazil). Anu Inst Geocienc 44:415362

Dowling RK (2011) Geotourism's global growth. Geoheritage 3(1):1–13

Dowling RK, Newsome D (2005a) Geotourism. Routledge, London, pp 1–260

Dowling RK, Newsome D (2005b) Geotourism's issues and challenges. In: Geotourism. Routledge, London, pp 242–254

Dowling RK, Newsome D (2010) Global geotourism perspectives. Goodfellow Publishers, Oxford

El Hamidy M, Errami E, Elkaichi A (2024) An overview of scientific research on geoheritage in Morocco. Proc Geol Assoc 135(2):162–180

Elliot S (2010) Tourism away from the mainstream: the Travel and Tourism Research Association Canada conference. Tourismos 5(2):225–231

Ellis N (2011) The geological conservation review (GCR) in Great Britain-Rationale and methods. Proc Geol Assoc 122(3):353–362

El-sharkawy OK (2015) Evaluating tourist satisfaction in accordance to tour guides performance: heritage guided tours in Egypt. Tourismos 10(1):81–99

Erikstad L (2008) History of geoconservation in Europe. Geol Soc Spec Pub 300:249–256

Erikstad L (2013) Geoheritage and geodiversity management - the questions for tomorrow. Proc Geol Assoc 124(4):713–719

Fassoulas C, Mouriki D, Dimitriou-Nikolakis P, Iliopoulos G (2012) Quantitative assessment of geotopes as an effective tool for geoheritage management. Geoheritage 4(3):177–193

Fepuleai A, Weber E, Németh K, Muliaina T, Iese V (2017) Eruption styles of samoan volcanoes represented in tattooing, language and cultural activities of the indigenous people. Geoheritage 9(3):395–411

Fepuleai A, Németh K, Muliaina T (2021) Geopark impact for the resilience of communities in Samoa, SW Pacific. Geoheritage 13(3):50

Ferdowsi S (2024) Key actors of geoheritage conservation in tourism destinations. Geoheritage 17(1):8

Fletcher SM, Thiessen J, Gero A, Rumsey M, Kuruppu N, Willetts J (2013) Traditional coping strategies and disaster response: examples from the south pacific region. J Environ Public Health 2013:264503

Flor-Blanco G, Adrados L, Llana-Fúnez S, Bruschi V (2023) The blowholes fields of Pría and Arenillas (Asturias). World geological heritage? Geogaceta 74:91–94

Fonseca Filho RE, Moreira JC (2017) Geotourist profile of Itacolomi State Park, Ouro Preto and Mariana (MG). Espacios 38(47):1–18

Fonseca Filho RE, Castro PTA, Varajão AFC, Figueiredo MA (2018) Perceptions of Cipó mountain range national park (MG, Brazil) visitors for geotourism. Anu Inst Geocienc 41(2):520–537

Fox N, Graham LJ, Eigenbrod F, Bullock JM, Parks KE (2020) Incorporating geodiversity in ecosystem service decisions. Ecosyst People 16(1):151–159

Frey ML, Schmitz P, Weber J (2021) Messel pit UNESCO world heritage fossil site in the UGGp Bergstraße-Odenwald, Germany–challenges of geoscience popularization in a complex geoheritage context. Geoconserv Res 4(2):524–546

Frisk EL, Volchko Y, Sandström OT, Söderqvist T, Ericsson LO, Mossmark F, Lindhe A, Blom G, Lång LO, Carlsson C, Norrman J (2022) The geosystem services concept – what is it and can it support subsurface planning? Ecosyst Serv 58:101493

Fuertes-Gutiérrez I, Fernández-Martínez E (2010) Geosites inventory in the Leon Province (Northwestern Spain): a tool to introduce geoheritage into regional environmental management. Geoheritage 2(1):57–75

Fung CKW, Jim CY (2015) Unraveling Hong Kong Geopark experience with visitor-employed photography method. Appl Geogr 62:301–313

Gibson D, Pratt S, Iaquinto BL (2022) Samoan perceptions of travel and tourism mobilities–the concept of Malaga. Tour Geogr 24(4–5):737–758

Goldsmith R, McCue K (2016) Revised hazard assessment for Afulilo Dam, Samoa. Geol Soc Eng Geol Spl Publ 27(1):169–177

Golfinopoulos V, Papadopoulou P, Koumoutsou E, Zouros N, Fassoulas C, Zelilidis A, Iliopoulos G (2022) Quantitative assessment of the geosites of Chelmos-Vouraikos UNESCO global geopark (Greece). Geosciences 12(2):63

Gomes CSF, Fernandes JV, Fernandes FV, Silva JBP (2021) Salt mineral water and thalassotherapy. In: Minerals latu sensu and human health: benefits, toxicity and pathologies. Springer, Cham, pp 631–656

Gonera M (2010) Geoconservation of landscape parks in Poland. Przegl Geol 58(12):1139–1146

Gordon JE (2019) Geoconservation principles and protected area management. Int J Geoheritage Parks 7(4):199–210

Gordon JE, Barron HF (2012) Valuing geodiversity and geoconservation: developing a more strategic ecosystem approach. Scott Geogr J 128(3-4):278–297

Gravis I, Németh K, Twemlow C, Németh B (2020) The ghosts of old volcanoes, a geoheritage trail concept for Eastern Coromandel Peninsula. N Z Geoconserv Res 3(1):40–57

Gray M (2008) Geodiversity: developing the paradigm. Proc Geol Assoc 119(3–4):287–298

Gray M (2011) Other nature: geodiversity and geosystem services. Environ Conserv 38(3):271–274

Gray M (2012) Valuing geodiversity in an 'ecosystem services' context. Scott Geogr J 128(3–4):177–194

Grigorescu DA (2020) From scientific research to geoconservation and geopark. Geoconserv Res 3(2):8–31

Guilbaud MN, Villalba NI, Osnaya ER, del Pilar Ortega-Larrocea M, Zeno-Lira E, Cram-Heydrich S (2024) The multiple values of urban geosites: El Arenal and Viaje a la Naturaleza parks in Mexico City. Geofis Int 63(3):1067–1085

Hall CM (1994) Is tourism still the plantation economy of the south pacific the case of Fiji. Tour Recreat Res 19(1):41–48

Harrison D (2008) Pro-poor tourism: a critique. Third World Q 29:851–868

Harrison D (2022) Mass tourism in small pacific island states. In: Stephenson ML (ed) Routledge handbook on tourism and small island states in the pacific. Routledge, London

Henriques MH, dos Reis RP, Brilha J, Mota T (2011) Geoconservation as an emerging geoscience. Geoheritage 3(2):117–128

Henriques MH, Tomaz C, Sá AA (2012) The Arouca Geopark (Portugal) as an educational resource: a case study. Episodes 35(4):481–488

Herrera-Franco GA, Carrión-Mero PC, Mora-Frank CV, Caicedo-Potosí JK (2020) Comparative analysis of methodologies for the evaluation of geosites in the context of the Santa Elena-Ancón geopark project. Int J Des Nat Ecodyn 15(2):183–188

Herrera-Franco G, Mata-Perelló J, Caicedo-Potosí J, Carrión-Mero P (2023) Vulnerability in geosites: a systematic literature review, World sustainability series. Springer, Cham, pp 395–407

Holmgaard SB (2019) The role of religion in local perceptions of disasters: the case of post-tsunami religious and social change in Samoa. Environ Hazards 18(4):311–325

Hose TA (2004) Geotourism - appreciating the deep time of landscapes. In: Niche tourism: contemporary issues, trends and cases. Elsevier, Cham, pp 27–38

Hose TA (2008) Towards a history of geotourism: definitions, antecedents and the future. Geol Soc Spec Pub 300:37–60

Hose TA (2012) 3G's for modern geotourism. Geoheritage 4(1–2):7–24

Huber ME (1994) An assessment of the status of the coral reefs of Papua New Guinea. Mar Pollut Bull 29(1–3):69–73
Hughes E (2022) Tourism and the sustainable development goals in small island states in the pacific. In: Stephenson ML (ed) Routledge handbook on tourism and small island states in the Pacific, 1st edn. Routledge, London
Jennings G (2007) Water-based tourism, sport, leisure, and recreation experiences. In: Water-based tourism, sport, leisure, and recreation experiences. Elsevier, Amsterdam, pp 1–20
Kathiresan K (2021) Mangroves: types and importance. In: Mangroves: ecology, biodiversity and management. Springer, Cham, pp 1–31
Kelly C (2012) Wellness tourism: retreat visitor motivations and experiences. Tour Recreat Res 37(3):205–213
Kiddell E (2012) Samoa's culture, tourism and development: exploring the impact of tourism development on the changing Fa'a Samoa. In: Lund University master of international development and management. Lund University, Lund, p 52
Kreisel W (1996) Tourism in developing countries - the example of the Pacific Island area. Geogr Schule 18(99):16–21
Kreisel W, Reeh T (2006) Tourism on the islands of the North Pacific. Geogr Rundsch 58(9):12–19
Kubalíková L, Balková M (2023) Two-level assessment of threats to geodiversity and geoheritage: a case study from Hády quarries (Brno, Czech Republic). Environ Impact Assess Rev 99:107024
Kubalíková L, Irapta PN, Pál M, Zwoliński Z, Coratza P, Vries BW (2023) Visages of geodiversity and geoheritage: a multidisciplinary approach to valuing, conserving and managing abiotic nature. Geol Soc Spec Pub 530(1):1–12
Kuchumov AV, Eremicheva PY, Bogrov IV (2024) Specifics of sustainable development of medical tourism in South Korea. In: Springer proceedings in business and economics. Springer, Cham, pp 451–463
Kumar RR, Stauvermann PJ (2023) Tourism and economic growth in the Pacific region: evidence from five small island economies. J Asia Pac Econ 28(3):894–921
La Russa MF, Patanè A, Apollaro C, Bloise A, Fuoco I, Ricca M, Russo L, Vespasiano G (2024) Quantitative assessment of geosites and mine heritage as a resource: the case study of Lungro Salt Mine (Calabria, Italy). Geoheritage 16(3):77
Lamb MP, Mackey BH, Farley KA (2014) Amphitheater-headed canyons formed by megaflooding at Malad Gorge, Idaho. Proc Natl Acad Sci USA 111(1):57–62
Larwood JG, Badman T, McKeever PJ (2013) The progress and future of geoconservation at a global level. Proc Geol Assoc 124(4):720–730
Lee KH, Lee TJ (2015) Opportunities and issues in the health tourism industry: deep sea water development in Taiwan. Tour Anal 20(4):419–424
Lilomaiava-Doktor S (2009) Beyond migration: Samoan population movement (malaga) and the geography of social space (Vā). Contemp Pac 21(1):1–32
Lima EA, Lima AF, Brilha J, Calado H, Nunes JC (2013) Integration of the geoheritage management in environmental and land-use policies. In: Rendiconti online societa geologica italiana. Soc Geol Ital, Rome, pp 102–105
Lima EA, Nunes JC, Costa MP, Machado M (2014) Azores Geopark: the challenge of managing an archipelagic Geopark. In: International multidisciplinary scientific GeoConference surveying geology and mining ecology management. SGEM, New York, pp 77–84
Liu S, Tian M, Shi W (2010) Digital Earth, a new approach for the geoconservation -case study of Hexigten Global Geopark, Inner Mongolia. In: Proceedings of SPIE - The International Society for Optical Engineering
Lopes MC, Liberato D, Alén E (2024) The influence of the senior tourist profile on the assessment of the dimensions of well-being and quality of life. In: Smart innovation, systems and technologies. Springer, Cham, pp 95–110
Lovelock B, Tuipoloa-Utuva L, Carr A (2022) Sustainable tourism planning in Samoa: identifying the challenges and opportunities. In: Routledge handbook on tourism and small island states in the pacific. Routledge, London, pp 313–327
Macpherson C, Macpherson L (2011) Churches and the economy of Samoa. Contemp Pac 23(2):304–337
Mahboubi CY, Naimi MN (2024) Geosites inventory and proposal of georoutes throughout El Bayadh Area (Saharan Atlas, Algeria). Geoheritage 16(3):1–13
Marie G (2003) The basaltic cliffs of Hawaï island: mode and rate of the retreat. Hommes Terres Nord 1:2–11
Mehdioui S, Hadi HE, Tahiri A, Haibi HE, Tahiri M, Zoraa N, Hamoud A (2022) The geoheritage of Northwestern Central Morocco area: inventory and quantitative assessment of geosites for geoconservation, geotourism, geopark purpose and the support of sustainable development. Geoheritage 14(3):86
Mehdipour Ghazi J, Audra P (2022) Travertine park in Azarshahr (NW Iran): an opportunity for geoheritage conservation and diminishing geohazards risk. Geoheritage 14(3):99
Migoń P (2024) Geosites and climate change—a review and conceptual framework. Geosciences 14(6):153
Migoń P, Pijet-Migoń E (2024) Geoconservation history of a basalt quarry – the case of Mt. Wilkołak, land of extinct volcanoes geopark, SW Poland. Geoheritage 16(3):65
Miklós L, Špinerová A, Belčáková I, Offertálerová M, Miklósová V (2020) Ecosystem services: the landscape-ecological base and examples. Sustain For 12(23):1–23
Mohanan M, Shekhar SK (2022) Bibliometric analysis of publications on wellness tourism. Tourism 32(2):69–86
Monge-Ganuzas M, Guillén-Mondéjar F, Díaz-Martínez E, Herrero N, Brilha J (2024) Geoconservation at the International Union for Conservation of Nature (IUCN). Philos Trans R Soc A Math Phys Eng Sci 382:2269
Moura P, Garcia MGM, Brilha JB, Amaral WS (2017) Conservation of geosites as a tool to protect geoheritage: the inventory of Ceará Central Domain, Borborema Province - NE/Brazil. Anais Acad Bras Cien 89(4):2625–2645
Movono A, Scheyvens R, Auckram S (2022) Silver linings around dark clouds: tourism, Covid-19 and a return to traditional values, villages and the vanua. Asia Pac Viewp 63(2):164–179
Naimi MN, Cherif A (2021) Inventory and assessment of significant scientific Algerian geoheritage: case of remarkable geosites from Orania (Western Algeria). Int J Geoheritage Parks 9(1):13–29
Németh K (2022) Geoheritage and geodiversity aspects of catastrophic volcanic eruptions: lessons from the 15th of January 2022 Hunga Tonga - Hunga Ha'apai eruption, SW Pacific. Int J Geoheritage Parks 10(4):546–568
Németh K, Gravis I (2022) Geoheritage and geodiversity elements of the SW Pacific: a conceptual framework. Int J Geoheritage Parks 10(4):523–545
Németh K, Fepuleai A, Muliaina T (2017) Samoa geopark project: a strategic plan to promote the volcanic geoheritage of Savai'i Island, Samoa. STAR 2017 - the pacific islands science, technology and resources conference [26–29 June 2017]. https://doi.org/10.13140/RG.2.2.19607.68009
Németh B, Németh K, Procter JN (2021a) Informed geoheritage conservation: determinant analysis based on bibliometric and sustainability indicators using ordination techniques. Land 10(5):539
Németh K, Gravis I, Németh B (2021b) Dilemma of geoconservation of monogenetic volcanic sites under fast urbanization and infrastructure developments with special relevance to the auckland volcanic field, New Zealand. Sustainability 13(12):6549
Newsome D, Dowling RK (2005) The scope and nature of geotourism. In: Geotourism. Routledge, London, pp 3–25

Nistor EL, Dezsi Ş (2022) An insight into gastronomic tourism through the literature published between 2012 and 2022. Sustain For 14(24):16954
Novelli M (2004) Niche tourism: contemporary issues, trends and cases. Elsevier, Amsterdam, pp 1–264
Novelli M, Benson A (2004) Niche tourism: a way forward to sustainability? In: Niche tourism: contemporary issues, trends and cases. Elsevier, Amsterdam, pp 247–250
Novković I, Dragićević S, Djurović M (2022) Geohazard and geoheritage. In: World regional geography book series. Springer, Cham, pp 119–131
Ólafsdóttir R (2019) Geotourism. In: Geosciences. Springer, Cham
Ólafsdóttir R, Dowling R (2014) Geotourism and geoparks-a tool for geoconservation and rural development in vulnerable environments: a case study from iceland. Geoheritage 6(1):71–87
Parkes MA (2008) A history of geoconservation in the Republic of Ireland. Geol Soc Spec Pub 300:237–248
Pasquaré Mariotto F, Drymoni K, Bonali FL, Tibaldi A, Corti N, Oppizzi P (2023) Geosite assessment and communication: a review. Resources 12(2):29
Pearce DG (2002) New Zealand holiday travel to Samoa: a distribution channels approach. J Travel Res 41(2):197–205
Piedade D, Guerreiro M, Pinto P (2024) Customer destination-based experience and loyalty: exploring the role of tourists' emotions. Consumer Behav Tour Hosp 19(4):478–496
Pimentel N, Pereira B, Silva MR (2024) From geoheritage to geosites at the oeste aspiring geopark (Portugal). Geoheritage 16(1):31
Piranha JM, Aparecida Del Lama E, de La Corte BD (2011) Geoparks in Brazil—strategy of geoconservation and development. Geoheritage 3(4):289–298
Poirine B (2010) The economy of French Polynesia: past, present and future. Pac Econ Bull 25(1):24–34
Pratt S, Harrison D (2015) Tourism in Pacific Islands: current issues and future challenges. Routledge, London, pp 1–298
Pratt S, Kuilamu M (2022) Tourism and the pacific island economies: outcomes and implications. In: Routledge handbook on tourism and small island states in the pacific. Routledge, London, pp 75–92
Prosser CD (2011) Principles and practice of geoconservation: lessons and case law arising from a legal challenge to site-based conservation on an eroding coast in Eastern England, UK. Geoheritage 3(4):277–287
Prosser CD (2012) William Archibald Macfadyen (1893-1985): the 'father of geoconservation'? Proc Geol Assoc 123(1):182–188
Prosser CD (2018) Geoconservation, quarrying and mining: opportunities and challenges illustrated through working in partnership with the mineral extraction Industry in England. Geoheritage 10(2):259–270
Quesada-Valverde ME, Quesada-Román A (2023) Worldwide trends in methods and resources promoting geoconservation, geotourism, and geoheritage. Geosciences 13(2):39
Quesada-Valverde M, Quesada-Román A (2025) Inventory and assessment of geosites to promote geotourism in Coto Brus. Costa Rica Geomorphol 470:109531
Rapaport M (2013) The pacific islands: environment and society. University of Hawaii Press, Hawaii, pp 1–459
Raposo VB, Melo CS, Silva L, Ventura A, Câmara R, Pombo J, Johnson ME, Ávila SP (2018) Comparing methods of evaluation of geosites: the fossiliferous outcrops of Santa Maria Island (Azores, NE Atlantic) as a case study for sustainable Island tourism. Sustain For 10(10):3596
Raymond EL (2023) The role of the diaspora: malaga, vā, and contesting the financialization of customary land in Samoa. Urban Geogr 44(2):295–297
Reynard E, Brilha J (2017) Geoheritage: assessment, protection, and management. Elsevier, Amsterdam, pp 1–450
Reynard E, Buckingham T, Martin S, Regolini G (2021) Geoheritage, geoconservation and geotourism in Switzerland. In: World geomorphological landscapes. Springer, Cham, pp 411–425
Risanti AA, Marfai MA (2020) The effects of hydrodynamic process and mangrove ecosystem on sedimentation rate in Kendal coastal area, Indonesia. IOP Conf Ser Earth Environ Sci 451:012070
Rivero MS, Rangel MCR, Martín JMS (2019) Geotourist profile identification using binary logit modeling: application to the Villuercas-Ibores-Jara Geopark (Spain). Geoheritage 11(4):1399–1412
Robinson M, Novelli M (2004) Niche tourism: an introduction. In: Niche tourism: contemporary issues, trends and cases. Elsevier, Amsterdam, pp 1–11
Robson AE (2009) Malietoa, Williams and Samoa's embrace of christianity. J Pac Hist 44(1):21–39
Rodrigues H, Brochado A, Troilo M, Mohsin A (2022) Wellness comes in salty water: thalassotherapy spas and service level of satisfaction. Int J Spa Wellness 5(1):71–90
Rogers K (2021) Accommodation space as a framework for assessing the response of mangroves to relative sea-level rise. Singap J Trop Geogr 42(2):163–183
Rosado-González EM, Palacio-Prieto JL, Sá AA (2020) Geotourism in Latin America and Caribbean UNESCO Global Geoparks: contribution for sustainable development goals. In: Studies on entrepreneurship, structural change and industrial dynamics. Springer, Cham, pp 107–121
Saintilan N, Rogers K (2013) The significance and vulnerability of Australian saltmarshes: implications for management in a changing climate. Mar Freshw Res 64(1):66–79
Sajinkumar KS (2013) Potential geohazards in Varkala coastal cliffs: an ideal hotspot for geotourism in South Western India. In: Rendiconti Online Societa Geologica Italiana. Societa Geologica Italiana, Rome, pp 129–132
Santos DS, Mansur KL (2024) The educational importance of geosites representing geodiversity-biodiversity relationships: a thematic inventory proposal. Geoheritage 16(4):99
Santos-González J, Marcos-Reguero A (2019) Applying the geological heritage in land management: cartography and management proposals of geosites in Burgos Province (Spain). Geoheritage 11(2):485–500
Schänzel H, Urie G, Lynch P (2022) Slow hospitality: family perspectives of holiday experiences at beach fale in Samoa. Hosp Soc 12(1):95–116
Scheyvens R (2005) Beach fale tourism in Samoa: the value of indigenous ownership and control over tourism
Scheyvens R, Movono A, Auckram J (2023) Enhanced wellbeing of Pacific Island peoples during the pandemic? A qualitative analysis using the Advanced Frangipani Framework. Int J Wellbeing 13(1):59–78
Seijmonsbergen AC, Sevink J, Cammeraat LH, Recharte J (2010) A potential geoconservation map of the Las Lagunas area, northern Peru. Environ Conserv 37(2):107–115
Simbaña-Tasiguano M, Cabascango-Chiliquinga E, Sánchez-Cortez JL, García-Frank A, Grefa-Shiguango H (2024) Geoeducation strategies in the Amazon, Napo Sumaco Aspiring UNESCO Global Geopark. Int J Geoheritage Parks 12(3):465–484
Somma R, Spoto SE, Giacobbe S (2024) Geological and structural framework, inventory, and quantitative assessment of geodiversity: the case study of the lake Faro and lake Ganzirri global geosites (Italy). Geosciences 14(9):236
Spyrou E, Maroukian H, Saitis G, Evelpidou N, Karkani A (2024) Promoting geo-education and geotourism through geosite assessment: a case study from Acheron-Parga, Epirus, Greece. Int J Geoheritage Parks 12(2):256–277
Sukojo BM, Annisa NN (2022) Analysis of sedimentation towards the changes of mangrove area using multitemporal remote sensing tech-

nology (a case study in gresik regency, East Java, Indonesia). Int J Adv Sci Eng Inf Technol 12(1):209–217

Ta'isi Efi TA (2018) An ethic of responsibility in Samoan customary law. In: Responsability: law and governance for living well with the earth. Routledge, London, pp 126–134

Tamang L, Mandal UK, Karmakar M, Banerjee M, Ghosh D (2023) Geomorphosite evaluation for geotourism development using geosite assessment model (GAM): a study from a Proterozoic terrain in eastern India. Int J Geoheritage Parks 11(1):82–99

Thornton A, Kerslake MT, Binns T (2010) Alienation and obligation: religion and social change in Samoa. Asia Pac Viewp 51(1):1–16

Thornton A, Sakai M, Hassall G (2012) Givers and governance: the potential of faith-based development in the Asia Pacific. Dev Pract 22(5-6):779–791

Tomić N, Božić S (2014) A modified geosite assessment model (M-GAM) and its application on the Lazar Canyon area (Serbia). Int J Environ Res 8(4):1041–1052

Trotman IG (1979) Western Samoa launches a National Park program. Parks 3(4):5–8

Turner S (2013) Geoheritage and geoparks: one (Australian) woman's point of view. Geoheritage 5(4):249–264

Valentine PS, Birtles A, Curnock M, Arnold P, Dunstan A (2004) Getting closer to whales - passenger expectations and experiences, and the management of swim with dwarf minke whale interactions in the Great Barrier Reef. Tour Manag 25(6):647–655

Vandelli V, Migoń P, Palmgren Y, Spyrou E, Saitis G, Andrikopoulou ME, Coratza P, Medjkane M, Prieto C, Kalovrektis K, Lissak C, Papadopoulos A, Papastamatiou N, Evelpidou N, Maquaire O, Psycharis S, Stroeven AP, Soldati M (2024) Towards enhanced understanding and experience of landforms, geohazards, and geoheritage through virtual reality technologies in education: lessons from the GeoVT project. Geosciences 14(5):127

Vasiljević DA, Vujičić MD, Bozić S, Jovanović T, Marković SB, Basarin B, Lukić T, Carkadcić J (2018) Trying to underline geotourist profile of National park visitors: case study of NP Fruška Gora, Serbia (Typology of potential geotourists at NP Fruška Gora). Open Geosci 10(1):222–233

Vilas Boas M, Brilha JBR, De Lima FF (2013) Conservation of geological heritage of Geopark Araripe (Brazil): framework, strategies and constraints. Bol Paranaense Geosci 70:156–165

Vong F (2016) Application of cultural tourist typology in a gaming destination – Macao. Curr Issue Tour 19(9):949–965

Vujičić MD, Vasiljević DA, Marković SB, Hose TA, Lukić T, Hadžić O, Janićević S (2011) Preliminary geosite assessment model (GAM) and its application on Fruška Gora Mountain, potential geotourism destination of Serbia. Acta Geogr Slov 51:303

Wang G, Ye W, Lv Y (2019) Loess geoheritage and geohazard protective measures at Luochuan Loess National Geopark in NW China. Geoheritage 11(3):1089–1100

Williams S, Griffiths J, Miville B, Romeo E, Leiofi M, O'Driscoll M, Iakopo M, Mulitalo S, Ting JC, Paulik R, Elley G (2021) An impacts-based flood decision support system for a tropical pacific island catchment with short warnings lead time. Water 13(23):3371

Worton GJ (2008) A historical perspective on local communities and geological conservation. Geol Soc Spec Pub 300:137–146

Wu L, Taniguchi K, Tajima Y (2021) Impact of climate change on flood hazard at airports on pacific islands: a case study of faleolo international airport, Samoa. J Disaster Res 16(3):351–362

Zakharovskyi V, Kósik S, Li B, Németh K (2023) Geosite determination based on geodiversity assessment utilizing the volcanic history of a near-sea-level explosive eruption-dominated volcanic island: Tūhua/Mayor Island, New Zealand. Geol Soc 530(1):127–140

Zakharovskyi V, Németh K, Gravis I, Twemlow C (2024) Geosite recognition based on qualitative-quantitative assessment in the light of core geological features of a mio-pliocene volcanic arc setting of the Coromandel Peninsula, New Zealand. Geoheritage 16(1):19

Zhang L, Guo ZH, Li ZY (2013) Carbon storage and carbon sink of mangrove wetland: research progress. Chin J Appl Ecol 24(4):1153–1159

Zhang L, Li T, Wang G, Kwang JS, Nittrouer JA, Fu X, Parker G (2020) How canyons evolve by incision into bedrock: Rainbow Canyon, Death Valley National Park, United States. Proc Natl Acad Sci USA 117(26):14730–14737

Zuo RG, Fang SM (2007) Design and realization of the geoheritage management information system based on MapGIS. Geol Bull China 26(2):223–230

Synthesis

5

This book presents a comprehensive study on estimating Samoa's geodiversity using a complex quantitative-qualitative method. This method calculates the density of geological-geomorphological features and their relative abundance on Earth, providing a spatial distribution that can identify geosites. The study's global relative abundance estimate offers a map comparable to other regions, indicating moderate geodiversity due to Samoa's relatively simple geological history.

Samoa's geology is largely shaped by 2 million years of ocean island basalt volcanism from hotspot activity, with typical shield volcano evolution followed by erosion and rejuvenation. Upolu shows significant erosion, resulting in a typical tropical landscape with subdued central and north-eastern segments, while Savai'i remains a younger, more pristine landscape dominated by a large shield volcano, dorsal rifts, and a major collapse scar.

Overall, Samoa's topography is modest with limited variations, marked by escarpments and volcanic collapse scars on Savai'i and deeply incised lava flows in Upolu. These morphological and geological features provide areas of higher geodiversity, creating a global map showing moderate geodiversity across the island.

This discussion covers how moderate geodiversity is interpreted in geoheritage and its relevance to geotourism. While global geodiversity estimation helps understand a location's position globally, it often misses local geological and geomorphological diversity. For Samoa, an island with limited area, recognizing these differences is crucial for identifying significant geosites. The island's young volcanic features and diverse coastal types are vital for local geotourism, geoconservation, and geoeducation. This book outlines a coastal classification tool and suggests further research into shallow marine regions, especially reef zones, to enhance geotourism and conservation efforts.

This work highlights Samoa's living culture and practices, which are deeply connected to the geology and geography of the land. These intangible (unable to be touched; not having physical presence) geoheritage elements are crucial for understanding human-land-ocean interactions or sense of place concept (da Silva 2019; García-Romero et al. 2024; Koupatsiaris and Drinia 2024; Sá and Rocha 2020; Said et al. 2021). In Samoa, cultural expressions like dance, tattooing, and oral traditions are integral to the island's geocultural integrity (Fepuleai et al. 2017).

There is debate on whether such intangible elements should be included in geodiversity estimates and its place in the broader heritage context (Ali and Abdulzahra 2022; Bruschi and Cendrero 2005; Caetano and Ponciano 2021; Goemaere et al. 2021; Halder and Sarda 2021; Henriques 2023; Henriques and Neto 2023; Pijet-Migoń and Migoń 2022; Qiu et al. 2022; Reynard and Giusti 2018; Shajahan et al. 2024; Van Der Ryn 2019). This book includes material elements like architectural places using local volcanic rocks in geodiversity calculations, showing how these additions make the geodiversity map of Samoa more complex and useful for locating geosites.

Geosites are essential for creating geoconservation programs, protected areas, and geotourism sites. The research shows that geodiversity estimates are sensitive to cultural additions, an important consideration for regions like Samoa seeking UNESCO Global Geopark status.

The study also discusses how far geodiversity should include geographical, cultural, or social aspects. Samoa's strong cultural connection to the land suggests that these geoheritage elements should not be separated. Despite changes since European contact, Samoa's village culture remains closely linked to its natural environment and can be considered part of its geoheritage.

The argument illustrates Samoa's potential contribution to geoheritage and geodiversity research. Samoa exempli-

K. Németh, V. Zakharovskyi, *Geodiversity of Samoa*, Geoheritage, Geoparks and Geotourism,
https://doi.org/10.1007/978-3-031-87220-4_5

fies a geologically and geographically distinct area where Polynesian society has coexisted symbiotically with the geosystem for over 3000 years. Daily life relies on understanding these hazards for survival. This complex relationship between humans and nature aligns with the concept of heritage from a human perspective and has a strong "geo" aspect.

This book explores the state of conservation in Samoa. Despite limited funding, Samoa strives to preserve regions for future use and promote sustainable development. While current projects do not specifically mention geoconservation, many align with its principles through outcomes like reports and publications. By integrating geoconservation concepts into existing ecological and biodiversity programs, Samoa could become a leader in this field. International bodies such as IUGS, ProGEO, and IUCN emphasize support for small island nations amid global changes like sea level rise and climate change, making targeted geoconservation efforts well-received in Samoa.

Properly managed geoconservation at the government level and active community participation can co-develop sustainable programs. These programs can expand geotourism beyond mainstream ideas, leading to an island-wide conservation network that benefits local communities.

On a small island with a growing population, sustainable use of natural resources is vital. Sustainable tourism and conservation strategies, including geotourism, are key but face geoethical challenges (Peppoloni et al. 2019; Peppoloni and Di Capua 2022a, b, c, d). Integrating geoethics into conservation, ecosystem services, and education is essential (Peppoloni and Di Capua 2021a, b, c, 2022a, b, c, d). Tourism and geotourism must address overtourism with proper management (Vagena 2021; Williams et al. 2020). These issues can be managed with a co-developed system involving Samoan communities, who adapt well due to their sociocultural structure.

References

Ali SH, Abdulzahra GN (2022) The role of intangible values for historic places in stimulating tourism investment (Ur City in Iraq-case study). Int J Des Nat Ecodyn 17(3):383–390

Bruschi VM, Cendrero A (2005) Geosite evaluation: can we measure intangible values. Il Quaternario 18:293–306

Caetano JMV, Ponciano LCMO (2021) Cultural geology, cultural biology, cultural taxonomy, and the intangible geoheritage as new strategies for geoconservation. Geoheritage 13(3):79

da Silva CM (2019) Geodiversity and sense of place: local identity geological elements in Portuguese municipal heraldry. Geoheritage 11(3):949–960

Fepuleai A, Weber E, Németh K, Muliaina T, Iese V (2017) Eruption styles of Samoan volcanoes represented in tattooing, language and cultural activities of the indigenous people. Geoheritage 9(3):395–411

García-Romero L, Marrero-Rodríguez N, Dóniz-Páez J, Peña-Alonso C, Pérez-Chacón Espino E, Da Silva CP (2024) Use and transformation of beaches as a tourism resource by promoters and managers in oceanic islands. A conflict for geoheritage conservation and social preferences in the Canary islands. Ocean Coast Manag 258:107378

Goemaere E, Millier C, Declercq PY, Fronteau G, Dreesen R (2021) Legends of the Ardennes Massif, a cross-border intangible geocultural heritage (Belgium, Luxemburg, France, Germany). Geoheritage 13(2):28

Halder S, Sarda R (2021) Promoting intangible cultural heritage (ICH) tourism: strategy for socioeconomic development of snake charmers (India) through geoeducation, geotourism and geoconservation. Int J Geoheritage Parks 9(2):212–232

Henriques MH (2023) Broadening frontiers in geoconservation: the concept of intangible geoheritage represented by the 1755 Lisbon earthquake. Geoheritage 15(2):57

Henriques MH, Neto K (2023) A geo-itinerary to foster sustainable tourism in West African Islands: storytelling the evolution of the ancient Cameroon Volcanic Line Coral Reefs. Sustainability (Switzerland) 15(24):16863

Koupatsiaris AA, Drinia H (2024) Expanding geoethics: interrelations with geoenvironmental education and sense of place. Sustainability (Switzerland) 16(5):1819

Peppoloni S, Di Capua G (2021a) Current definition and vision of geoethics. In: Bohle M, Marone E (eds) Geo-societal narratives: contextualising geosciences. Springer International Publishing, Cham, pp 17–28

Peppoloni S, Di Capua G (2021b) Geoethics as global ethics to face grand challenges for humanity. Geol Soc Lond Spec Publ 508(1):13–29

Peppoloni S, Di Capua G (2021c) Geoethics to start up a pedagogical and political path towards future sustainable societies. Sustainability 13:10024

Peppoloni S, Di Capua G (2022a) Ethical problems and dilemmas in the geosciences. In: Peppoloni S, Di Capua G (eds) Geoethics: manifesto for an ethics of responsibility towards the earth. Springer International Publishing, Cham, pp 55–61

Peppoloni S, Di Capua G (2022b) From ethics to geoethics. In: Peppoloni S, Di Capua G (eds) Geoethics: manifesto for an ethics of responsibility towards the earth. Springer International Publishing, Cham, pp 19–29

Peppoloni S, Di Capua G (2022c) Geoethics and anthropogenic global changes. In: Peppoloni S, Di Capua G (eds) Geoethics: manifesto for an ethics of responsibility towards the earth. Springer International Publishing, Cham, pp 83–106

Peppoloni S, Di Capua G (2022d) Geoethics for an ecological humanism. In: Peppoloni S, Di Capua G (eds) Geoethics: manifesto for an ethics of responsibility towards the earth. Springer International Publishing, Cham, pp 107–123

Peppoloni S, Bilham N, Di Capua G (2019) Contemporary geoethics within the geosciences. In: Bohle M (ed) Exploring geoethics: ethical implications, societal contexts, and professional obligations of the geosciences. Springer International Publishing, Cham, pp 25–70

Pijet-Migoń E, Migoń P (2022) Geoheritage and cultural heritage—a review of recurrent and interlinked themes. Geosciences (Switzerland) 12(2):98

Qiu Q, Zuo Y, Zhang M (2022) Intangible cultural heritage in tourism: research review and investigation of future agenda. Land 11(1):139

Reynard E, Giusti C (2018) Chapter 8—The landscape and the cultural value of geoheritage. In: Reynard E, Brilha J (eds) Geoheritage. Elsevier, Amsterdam, pp 147–166

Sá AA, Rocha D (2020) Arouca UNESCO Global Geopark: geomorphological diversity fosters local development. In: World geomorphological landscapes. Springer, Cham, pp 329–340

Said MZ, Komoo I, Mohamad ET, Ali CA, Ahmad N, Wahid MEA, Rajimin MF (2021) Geological, biological, cultural and local wisdom heritage a key element of mersing geopark development. Bull Geol Soc Malaysia 71:89–98

Shajahan R, van Wyk de Vries B, Zanella E, Harris A (2024) Creating a sense of intangible science: making it understandable to a broad public via geoheritage. Int J Geoheritage Parks 12(3):396–415

Vagena A (2021) OVERTOURISM: definition and impact. Academia Letters. https://doi.org/10.20935/AL1207

Van Der Ryn M (2019) Visual anthropology and intangible cultural heritage preservation in American Samoa. Journal de la Societe des Oceanistes 148(1):145–155

Williams MA, McHenry MT, Boothroyd A (2020) Geoconservation and geotourism: challenges and unifying themes. Geoheritage 12(3):63